SpringerBriefs in Complexity

Springer Complexity

Springer Complexity is an interdisciplinary program publishing the best research and academic-level teaching on both fundamental and applied aspects of complex systems—cutting across all traditional disciplines of the natural and life sciences, engineering, economics, medicine, neuroscience, social and computer science.

Complex Systems are systems that comprise many interacting parts with the ability to generate a new quality of macroscopic collective behavior the manifestations of which are the spontaneous formation of distinctive temporal, spatial or functional structures. Models of such systems can be successfully mapped onto quite diverse "real-life" situations like the climate, the coherent emission of light from lasers, chemical reaction-diffusion systems, biological cellular networks, the dynamics of stock markets and of the internet, earthquake statistics and prediction, freeway traffic, the human brain, or the formation of opinions in social systems, to name just some of the popular applications.

Although their scope and methodologies overlap somewhat, one can distinguish the following main concepts and tools: self-organization, nonlinear dynamics, synergetics, turbulence, dynamical systems, catastrophes, instabilities, stochastic processes, chaos, graphs and networks, cellular automata, adaptive systems, genetic algorithms and computational intelligence.

The three major book publication platforms of the Springer Complexity program are the monograph series "Understanding Complex Systems" focusing on the various applications of complexity, the "Springer Series in Synergetics", which is devoted to the quantitative theoretical and methodological foundations, and the "SpringerBriefs in Complexity" which are concise and topical working reports, case-studies, surveys, essays and lecture notes of relevance to the field. In addition to the books in these two core series, the program also incorporates individual titles ranging from textbooks to major reference works.

More information about this series at http://www.springer.com/series/8907

Alexander Tarvid

Agent-Based Modelling of Social Networks in Labour–Education Market System

Springer

Alexander Tarvid
Faculty of Economics and Management
University of Latvia
Riga, Latvia

Riga Business School
Riga Technical University
Riga, Latvia

ISSN 2191-5326 ISSN 2191-5334 (electronic)
SpringerBriefs in Complexity
ISBN 978-3-319-26537-7 ISBN 978-3-319-26539-1 (eBook)
DOI 10.1007/978-3-319-26539-1

Library of Congress Control Number: 2015956179

Springer Cham Heidelberg New York Dordrecht London

Printed on acid-free paper

Springer International Publishing AG Switzerland is part of Springer Science+Business Media (www.springer.com)

Preface

Most of us would like to have a job that would support both physical well-being through providing enough income and psychological well-being reflected in high satisfaction with life. Unfortunately, not all jobs are like that. A sizeable part of the society considers their jobs as inappropriate for different reasons, starting from severe working conditions and ending with the acquired education and/or skills not being used in the job. This has a negative impact on their physical and psychological well-being. Another part of society is not employed at all, either seeking job or being inactive, which typically doesn't make one's life comfortable.

Not only we as individuals are interested in the jobs we find appropriate for ourselves. Whether most of us are able to find them is also of relevance to firms and policy-makers. Firms are increasingly caring about the level of job satisfaction of their employees, because it affects the productivity of the latter and, ultimately, their decision to stay or quit the job. Policy-makers care about it, because poor quality of employment, meaning large rates of employee–job mismatch, unemployment or inactivity, increases their chances to fail re-elections and lose their jobs and, in extreme cases, leads to civil unrest.

Among the most important prerequisites of successful employment are knowledge and skills, which come from education and experience. Using the language of economists, the most efficient allocation of individuals across jobs is when the characteristics of the former match those required in the latter. Otherwise, the individual is mismatched by having more or less education and/or skills than the job requires. Research showed that working in a job where the acquired education or skills are not used has considerable negative effects on both physical well-being (e.g. lower salaries and slower career) and psychological well-being (e.g. higher risk of depression and lower job satisfaction).

On a macro level, this result brings attention to the coordination between the education market and the labour market. Lack of such coordination was repeatedly noted by many, including the World Economic Forum [296], which classified structural unemployment caused by education and skills mismatch as a global threat.

This coordination may be improved in three ways, depending on which of the two markets is viewed as more important and which as the source of the problem.

One way is to blame the education system that is preparing too few graduates in some fields of study and too many in others. Those who hold this view propose to restructure the education system so that it produces only the graduates demanded by employers [56, 128]. Another way is to blame the labour market, which is unable to take advantage of the existing supply of graduates. If this is the case, it is proposed to restructure the labour market so that the country benefits from the specialists it has [199]. Still another way is to take a broader view and understand the reasons of the existing imbalances, which might well exist because of misalignment of education and labour markets. Then these issues should be addressed and the links between the two markets should be improved [53]. This would lead to both markets working more closely together and being better adapted to current and future developments. Clearly, this third way appears to be the most beneficial.

This leads to the necessity of considering the labour market together with the education market as one system or *labour–education market system* (LEMS). Education-related decisions of individuals affect their labour-market options and outcomes, and labour-market outcomes of individuals affect education-related decisions of other individuals. This is the core of a typical LEMS model.

The idea that education and labour markets should be modelled together is, of course, not new, and economists have frequently embodied education characteristics of individuals into labour-market models. Usually, economists use mathematical modelling techniques and, hence, have to ignore the effects on individuals' decisions from social networks for mathematical tractability. At the same time, there is a vast body of empirical evidence confirming that social networks are an important source of information for individuals and influence their decisions in both education and labour markets. Should we really ignore this important element in our models of LEMS? Certainly not in all cases.

Agent-based modelling allows to overcome not only the inability of social network modelling by standard mathematical economics techniques but also other shortcomings of standard economic modelling. This book is about using agent-based simulations to model LEMS with embedded social networks—more concretely, individual behaviour in LEMS where individual decisions are affected by social networks.

The book is written as a guide to using agent-based modelling for this purpose. It does not contain fully developed models—so if you're interested in such examples, you'll have to read the relevant articles, and this book references many of them. Rather, it contains a set of proposals on how different aspects of LEMS models should be constructed and an analysis of the approaches to their construction in the available literature. It also does not contain examples of programming code in a concrete agent-based modelling platform—programme listings that are included are written in pseudocode, so that you can easily implement them in your preferred platform. It is also not an encyclopaedia on individual behaviour (it would be much thicker if it were)—but it is a collection of facts and analyses that you should know about and consider while building your agent-based model of LEMS.

The book does not require you to be an expert in the field of agent-based modelling, education market, labour market or social networks (even if you are,

I hope you'll still find the book useful). I also hope that it will be interesting to a wide readership, to both undergraduate students and experienced researchers.

There are three main chapters in the book. Chapter 1 discusses the facts we know about individual behaviour and the role of social networks in LEMS from empirical literature, also noting theoretical support, where relevant. Then Chap. 2 substantiates the need to apply agent-based modelling to studying LEMS, discussing the benefits and drawbacks of this modelling method, and provides a step-by-step guide to it. Finally, Chap. 3 analyses how three large blocks of LEMS—education market, labour market and social networks—can be constructed in an agent-based model, based on existing literature and the empirical results discussed in Chap. 1.

I wish you a pleasant reading.

Riga, Latvia Alexander Tarvid

Contents

Chapter 1
Social Networks and Labour–Education Market System

Educational choice is an outcome deeply intertwined with prior social choices negotiated through structures of constraints and possibilities.
Valentini Moniarou-Papaconstantinou et al. [192, p. 323]

Two facts about human beings are widely accepted: they are social creatures and they behave in a bounded rational way. In particular, this results in substantial use of social networks in individual decision-making. Before dealing with the issues of modelling individual behaviour in the labour–education market system, we have to recall some empirical facts known from the literature about this behaviour. This is exactly what this chapter provides.

It starts with a discussion on what a social network is, how it is created and maintained, and what its typical structure is. Then I move to describing individual behaviour in the education market, focusing on the factors that are important in choosing the education path. The last topic of the chapter is individual behaviour in the labour market.

1.1 Social Networks

I was recently talking with a high official of a local higher education institution. He mentioned that they had just finished the analysis of how students used social networks while deciding whether to enrol there. I immediately asked whether I could get access to these results to use them in calibrating my agent-based models of LEMS. Unfortunately, it became clear shortly that they were doing nothing more than analysing how potential students used accounts of the institution on Facebook and in other online social network services, so I had to explain what's the difference between that and the social networks I was modelling.

Indeed, when someone refers to social networks nowadays, many understand that he or she refers to online social network services (websites and mobile applications) such as Facebook, Twitter or LinkedIn that allow to chat and share information with friends or colleagues. While important and quite pervasive, these are nothing more than a medium of exchange of information.

A. Tarvid, *Agent-Based Modelling of Social Networks in Labour–Education Market System*, SpringerBriefs in Complexity, DOI 10.1007/978-3-319-26539-1_1

In this book, I look on social networks in their broadest sense, which is a (mathematical) representation of the connections between individuals. Naturally, this representation is a graph where individuals are vertices, some of which are connected.

This seeming simplicity shouldn't mislead you. There are many types of social networks, and in different circumstances, individuals use different networks. Actually, one may view all of them as a collection of sets of edges or arcs superimposed on the same set of vertices.

This section is about the main principles individuals use to create and maintain their social networks. However, because you don't necessarily know the terminology used when discussing networks, I'll start with a concise introduction to the most important definitions from the graph theory that will be used in the book.

1.1.1 Required Minimum from the Graph Theory

A *graph* is an ordered pair $G = (V, E)$, where V is a set of vertices and E is a collection of edges. Informally speaking, every edge connects two vertices and, thus, may be represented as a pair of vertices. Different restrictions on these pairs may be introduced, and depending on these restrictions, different types of graphs are distinguished.

In this text, the only types of graphs of interest will be *undirected* and *directed* graphs, and I'll further mostly focus on undirected graphs. In an undirected graph, an edge is an unordered pair of vertices: pair (v, w) represents the same edge as pair (w, v). Some authors, thus, write edges in undirected graphs as sets, $\{v, w\}$, and I'll use this notation throughout this text. In a directed graph, edges are sometimes called *arcs* to distinguish them from those in undirected graphs. An arc (in a directed graph) is an ordered pair of vertices: arc (v, w) is not the same as arc (w, v), because they have opposite directions. In graph theory, the term *network* is sometimes used to refer only to directed graphs (in the context of different flows along arcs in the graph); in this book, I will use the term *network* as a synonym of the term *graph* and use it to refer to both directed and undirected graphs.

I'll further consider only graphs where there can be only zero or one edge or arc (with same direction) between the two given vertices. Note that it allows directed graphs to contain arcs (v, w) and (w, v) at the same time, as they have different directions. Thus, E in this text refers to a set, not just a collection, of edges/arcs, as it doesn't allow several copies of an edge/arc.

Another restriction is that *self-loops*, or edges/arcs connecting a vertex with itself, are not allowed in graphs. One could argue how relevant the existence or non-existence of loops is to the practical application of graphs to social networks in economics, but denying the possibility of a loop certainly simplifies the analysis and programming.

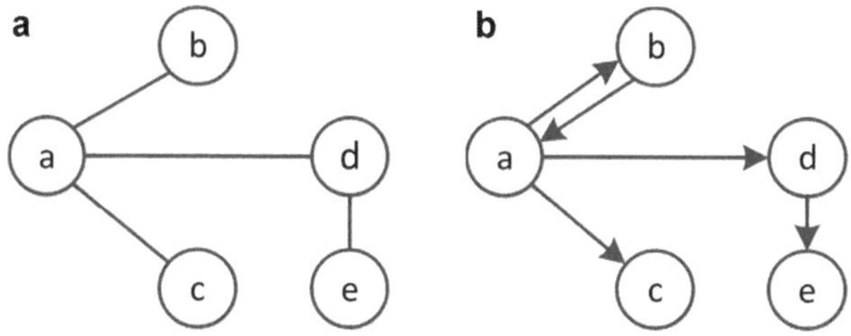

Fig. 1.1 (**a**) Example: undirected graph. Vertices: $V = \{a, b, c, d, e\}$, edges: $E = \{\{a, b\}, \{a, c\}, \{a, d\}, \{d, e\}\}$, degrees: $\deg a = 3, \deg d = 2, \deg b = \deg c = \deg e = 1$. (**b**) Example: directed graph. Vertices: $V = \{a, b, c, d, e\}$, arcs: $E = \{(a, b), (b, a), (a, c), (a, d), (d, e)\}$, out-degrees: $\deg^+ a = 3, \deg^+ b = \deg^+ d = 1, \deg^+ c = \deg^+ e = 0$, in-degrees: $\deg^- v = 1 \ \forall v \in V$

Consider edge $e = \{u, v\}$. Edge e is said to be *incident* to vertices u and v (we can also say that u and v are incident to e). Vertices u and v are said to be connected by edge e or *adjacent* (to each other).[1]

For a vertex, an important characteristic is the number of edges or arcs it is incident to. In an undirected graph, this is called a *degree* of vertex and formally defined as $\deg v = |\{w : \{v, w\} \in E\}|$. In a directed graph, vertices have two degrees: an *in-degree*, the number of incoming arcs, and an *out-degree*, the number of outgoing arcs. Formally, the in-degree of vertex v is $\deg^- v = |\{w : (w, v) \in E\}|$ and its out-degree is $\deg^+ v = |\{w : (v, w) \in E\}|$. Figures 1.1a, b provide illustrative examples.

A *path* is a sequence of *different* vertices (although the first and last vertices can be the same, in which case the path is called a *cycle*), where each two neighbouring vertices are connected by an edge. In directed graphs, the corresponding notion is *directed path*, where each two neighbouring vertices are connected by an arc so that the direction of the arc is always from the preceding vertex in the list to the next and not vice versa. The *length of the path* is the number of edges/arcs there. For instance, in Fig. 1.1a, path *ade* has length two. In Fig. 1.1b, directed path *ade* also has length two, but there is no directed path *eda*.

1.1.2 Creating, Maintaining and Dissolving Social Ties

Individuals make two main decisions regarding their social network: on its composition and size.

The first of them answers the question “Whom should I make friends with?” and a related question “Who should I drop from my friendship in the first place?”

[1]From these two terms follow two kinds of matrix representations of a graph: with an *incidence matrix* ($M_{ij}^{\text{inc}} = 1$ if vertex i and edge j are incident) and with an *adjacency matrix* ($M_{ij}^{\text{adj}} = 1$ if vertex i and vertex j are adjacent).

Theoretically, the answer to both questions is captured by the notions of homophily and assortative mating, which roughly mean the same thing (except that homophily is more general): *homophily* is "the principle that a contact between similar people occurs at a higher rate than among dissimilar people" [188, p. 416], while *assortative mating* is "preference for sexual partners with similar phenotypes, as when humans select mates who tend to resemble themselves in wealth, social class, intelligence, or attractiveness, or when other organisms choose mates that resemble themselves on specific criteria" [68]. Thus, ties tend to be created between more similar individuals and dissolved in the first place between less similar individuals.

Because of homophily, personal networks tend to be quite homogeneous on a large number of parameters characterising the individual. Moreover, homophily is found in many types of personal networks ranging from marriage to friendship to simply knowing someone. As shown in [188], such segregation of society is going by socio-demographic dimensions and at the level of values, beliefs and attitudes, which appear in the following order of decreasing importance:

1. Race and race-like ethnicity
2. Sex, age, religion and level of education
3. Occupation, network position, behaviours and intra-personal values

Homophily is not the only mechanism that affects the composition of the social network. The second mechanism is called *structural constraint* in [111]. The idea is that whatever the preference for network connections, the pool of potential members of the network is to a large extent determined (hence, constrained) by the social contexts in which the individual participates and, thus, is as such relatively homogeneous.

The above said does not mean that the social network is composed *exclusively* of members strongly similar to each other. The *strength of weak ties hypothesis* [122] postulates that if all the connections of an individual are *strong ties* in that they tend to use the same information pool as he uses then he will not receive novel information. *Weak ties*, in contrast, are moving in dissimilar information circles and, thus, allow to extract novel, potentially useful information. Hence their strength.

Ties are created in different contexts. For instance, the left panel of Fig. 1.2 presents a distribution of social network ties by the context in which they were created. Most of them (around 40 %) are created through family, as compared to, for instance, only 10 % created during education. Individual's education affects the distribution of contexts where ties are created: the share of ties created during education, at work or in organisations increases tremendously with education level [125]. Ties created in one context do not necessarily remain in that context, however. For instance, a new acquaintance from the university may become a friend or a person met through friends may become part of the family or, alternatively, the relationship can be broken. The right panel of Fig. 1.2 presents a general composition of a social network. It shows that the three largest groups of ties are family-related ties (around 40 % of all ties), friendship ties (around 30 %) and ties with co-workers and organisational members (around 20 %).

The second decision on the social network answers the question "How many social ties should I keep?" Individuals cannot have as many social ties as they

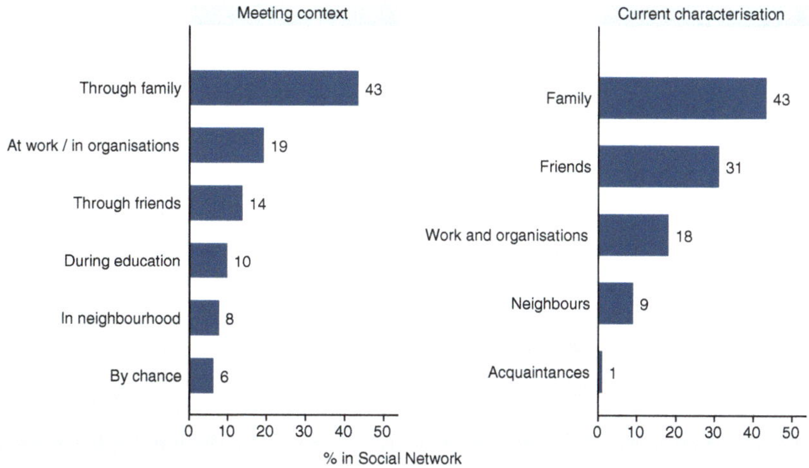

Fig. 1.2 Social network structure, by meeting context and current characterisation of ties, Toulouse area, France. *Source:* developed based on [125]. The figure allows to compare the frequency of tie creation in different contexts to how the ties are characterised some time after creation, at the moment of surveying. The data on the current characterisation of social ties are roughly consistent with US-based data from [111], where 42 % of ties are family-related, 23 % are friends, 16 % are co-workers or organisation members, 10 % are neighbours and 6 % are "others"

would perhaps like to because maintaining each social tie has a cost, which might be cognitive, emotional or temporal, as well as because there are spatial limits on the number of ties they can sustain [122]. Even today, when due to the abundance of online social network services, spatial limits are significantly reduced, the three costs mentioned above still remain. Based on the size of human neocortex, it was predicted that the average size of social networks should be around 150 [99], which was later supported for a network of Christmas card exchange [136] and for the number of connections on Twitter [120]. A systematic analysis of empirical studies on social network sizes concluded that networks are built hierarchically, each next hierarchical level being 3–4 times larger than the previous one: support clique level (3–5 individuals), sympathy level (9–15), band level (30–50) and community level (around 150) [299].[2] The *support clique*,[3] formed by the closest individuals, from whom personal advice or help in severe circumstances is sought, is the core and most frequently contacted component of the network. The frequency of contact

[2]There are also two larger groupings: mega-band level (around 500 individuals) and large tribe level (around 1500 individuals) [99].

[3]A *clique* is a set of vertices where each vertex is connected with all other vertices from the set.

decreases with hierarchy level. The hierarchical structure is also observed in online social networks, and it is similar in terms of sizes and contact frequencies to offline networks [100].

1.1.3 Structure of Social Networks

A particularity of social networks is that they have many more vertices with low degrees than with high degrees, which also makes sense because, as noted above, keeping each additional social tie entails a cost on the individual. Several structures that have this property were identified.

Many complex networks tend to be *scale-free*, meaning that degrees in these networks are distributed according to the *power law*. This simply means that the probability that vertex v has degree k decreases with k according to the following formula:

$$\Pr(\deg v = k) = ck^{-\beta}, \tag{1.1}$$

where c is a normalisation constant[4] and, typically, $\beta \in (2, 3)$ [23]. Studies found that degree distribution indeed follows power law in such online networks as Flickr and Yahoo! 360 [163], which are mainly used for sharing photos with friends; Twitter, a micro-blogging service, but only below 10^5 followers[5] [164]; Cyworld, a popular social network service in South Korea, but only for degrees above 10^3 [5]; and Facebook for both the nominal network declared there and the network with which users actually interact [294].

Two other degree distributions were identified in social networks. Degrees are distributed exponentially in the already-mentioned Cyworld social network for degrees below 10^3 [5], Facebook student network at Princeton University [7] and in networks of email addressees [2, 127]. Exponential degree distribution may be written as

$$\Pr(\deg v = k) = ce^{-\lambda k}, \tag{1.2}$$

where c is a normalisation constant and λ is a parameter.

Log-normal distribution of degrees was found in a mutual friendship network on a Finnish web service [273]. By definition, a random variable X is distributed log-normally with parameters μ and σ if $\ln X$ is distributed normally with the same parameters. Formally, the probability density function of a log-normal distribution is

$$f(x, \mu, \sigma) = \frac{1}{\sqrt{2\pi}\sigma x} e^{-(\ln x-\mu)^2/2\sigma^2}, \tag{1.3}$$

[4]The constant ensures that probabilities sum to one.

[5]The departure of degree distribution from power law for degrees above 10^5 was attributed to a large number of celebrities and large corporations using Twitter.

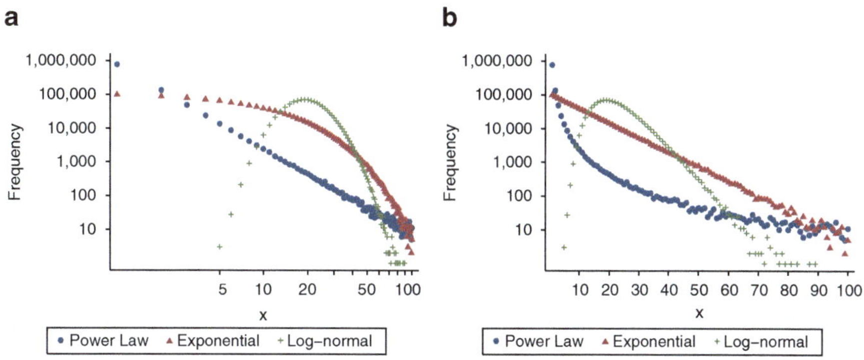

Fig. 1.3 (**a**) Distributions in log–log scale. (**b**) Distributions in log–lin scale. Distribution parameters: power law's exponent $\beta = 2.5$, exponential distribution's $\lambda = 0.10$, log-normal distribution's $\mu = 3, \sigma = 0.3$

although the distribution of degrees is, obviously, discrete, rather than continuous.

Examples of scale-free, exponential and log-normal distributions are shown in Fig. 1.3a, b. Visually, it's easy to distinguish a power law distribution from an exponential distribution: the former is a straight line in the log–log scale (and U-shaped in the log–lin scale), while the latter is linear in the log–lin scale (and inverse U-shaped in the log–log scale). Log-normal distribution is somewhat different, looking as a straight line in *both* log–log (thus, looking similarly to a power law[6]) and log–lin scales for most of its support.

Besides degree distribution, two other important characteristics of a social network are average degree and transitivity. While the former is clear, the latter requires explanation. *Transitivity* is defined by *clustering coefficient*, which ranges from 0 to 1 and is given by the following formula:

$$C = \frac{(\text{number of triangles}) \times 6}{(\text{number of paths of length two})}. \tag{1.4}$$

In words, it shows the share of cases when if it is known that v is connected to w and w is connected to u then also v is connected to u. The factor of 6 appears under the assumption that each triangle in the graph is counted once (and not six times: as vwu, vuw, wvu, wuv, uvw and uwv) and each path of length two is counted twice (once as vwu and again as uwv).

An alternative definition of a clustering coefficient exists and is frequently used in the literature, which creates certain confusion. It starts with defining a *local clustering coefficient* of vertex v,

$$C_v = \frac{|\{(u, w) : \{v, u\} \in E \wedge \{v, w\} \in E \wedge \{u, w\} \in E\}|}{\frac{1}{2} \deg v (\deg v - 1)}, \tag{1.5}$$

[6]See [190] for a lengthier treatment on the similarities between power law and log-normal distributions.

Table 1.1 Clustering coefficient and average clustering coefficient values in real data

Network	C	$\overline{C}$	Source
BibSonomy (social bookmarking system), friend connections	0.28		[191]
Cyworld		0.16	[5]
E-mail network, University at Rovira i Virgili		0.25	[127]
Facebook	0.16		[7, 294]
Facebook, users who downloaded custom application	0.57		[50]
Flickr, contact connections	0.05		[191]
last.fm, mutual friendships		0.31	[273]
Microsoft Messenger communication activities		0.14	[170]
MySpace		0.26	[5]
Orkut		0.31	[5]
Twitter, re-tweet connections	0.06		[191]
Twitter, follower connections	0.01		[191]
Whisper, communication activities		0.03	[291]
Wretch (blog-based social network, Taiwan)	0.10		[65]

that is, the probability that two neighbours of v are connected. The *average clustering coefficient* is then a simple average of local clustering coefficients:

$$\overline{C} = \frac{1}{|V|} \sum_{v \in V} C_v \, . \tag{1.6}$$

The confusion arises from some authors referring to C as transitivity and to $\overline{C}$ as clustering coefficient (skipping the word *average*).

As Table 1.1 shows, both clustering coefficient C and average clustering coefficient $\overline{C}$ in social networks typically are in the range between 0.10 and 0.30, although smaller and larger values are also possible.

1.2 Individual Behaviour in the Education Market

Recently I heard a conversation between two fresh secondary school graduates. One of them was bragging that he finished prestige programming courses and was now planning to continue studies at the faculty of computer science of a local university. Then he asked his friend about the field of study he was going to choose. "Well, I'm not sure yet," the friend replied. "Maybe, something like economics." The wannabe programmer made a wry smile in reply and noted that he didn't know any graduate of economics who then worked as an economist. His friend made another attempt: "Actually, I was looking very much forward to political science." The reaction he got in reply was a laughing "D'you really wanna be a politician? So boring! And there are too many of them already. Better choose something else." The undecided friend looked a bit frustrated and said "Maybe.. So where's your computer science faculty located?"

Some of the choices individuals make in the education market are: Should I remain with secondary education or get higher education? Should I choose a vocational track or an academic track? Which education institution should I go to? Which field of study should I choose?

In making many of these choices, they are guided by three groups of factors: their idiosyncrasies such as personality and ability, social environment and expected monetary returns. The importance a particular person assigns to each of these groups is different. For some, social outcomes such as prestige or recognition are more important, others put more weight on salary and career growth, while still others just simply want to realise their dream even if it has moderate social and monetary returns.

1.2.1 Individual Aspects: Ability, Sex and Personality

Research shows that individual's performance on key skills related to a particular field of study increases the willingness to choose that field and to succeed in it [274, 282]. For instance, students good at reading tend more to choose social sciences and humanities, while those good at mathematics tend to study natural science and engineering.

Sex is another important factor in choosing education path. For several reasons, including culture and tradition, the division into "female" fields (e.g., arts, humanities or education) and "male" fields (e.g., science or technology) still exists, which is very well documented in various countries [40, 55, 79, 121, 210, 282].

Not only are particular skills important for a given job, but also certain personality traits are important for success. On a theoretical level, this relationship is explained by the *personality–job fit theory*, which argues that individuals are more satisfied with the jobs whose types are congruent with their personalities.

The most well-known and tested concept in this theory is Holland's six types of personalities and environments [141]. He distinguishes realistic, investigative, artistic, social, enterprising and conventional personalities. Then, he argues, these could be put in the edges of a hexagon (in the order listed above) to show which types correspond better to each other. For instance, Realistic persons feel most comfortable in Realistic environments (zero distance between vertices) and least comfortable in Social environments (distance is two vertices). Table 1.2 summarises the characteristics of Holland's personality types and lists the corresponding fields of study.

Currently, Holland's six types are part of Strong Interest Inventory under the name General Occupational Themes. However, because it considers only six types, its accuracy in classifying individuals into fields of study or occupations is not as strong as one would like it to be. Thus, other classifications, e.g., Basic Interest

Table 1.2 Holland's personality types and related fields of study

Realistic personality:

- Prefer activities involving the manipulation of materials, tools or machines
- Avoid educational and interpersonal activities
- Practical, conservative, persistent, conforming, shy
- Lack social and educational competencies
- Value material rewards

Typical fields of study: Electrical/mechanical engineering, military/marine science

Investigative personality:

- Prefer exploration, understanding and prediction
- Avoid persuasive, social and repetitive activities
- Analytical, critical, independent, precise, rational
- Lack persuasive and leadership competencies
- Value knowledge acquisition and scholarly achievements in science and technology

Typical fields of study: Biology, natural science, economics, finance

Artistic personality:

- Prefer ambiguous, free and non-systematised activities to create art forms of products
- Avoid routine and conformity to rules
- Expressive, original, intuitive, independent, emotional, sensitive
- Lack clerical and business system competencies
- Value aesthetic qualities and creative expression of ideas, emotions or sentiments

Typical fields of study: Fine arts, languages, architecture

Social personality:

- Prefer to inform, teach, cure or enlighten others
- Avoid explicit, systematic activities involving materials, tools or machines
- Cooperative, empathetic, generous, helpful, sociable, possessing leadership competency
- Are not mechanically inclined
- Value education and social service

Typical fields of study: Humanities, nursing, psychology, political science

Enterprising personality:

- Prefer persuading and directing others to attain organisational goals or economic gain
- Avoid scientific, intellectual and abstruse activities
- Aggressive, ambitious, energetic, have leadership ability, self-confident
- Lack scientific ability
- Value material accomplishment and social status

Typical fields of study: Business, management, journalism, computer science

Conventional personality:

- Prefer explicit, ordered and systematic manipulation of data
- Avoid ambiguous and unstructured activities
- Careful, conforming, orderly, have clerical and numerical ability, efficient, inflexible
- Lack artistic competencies
- Value material accomplishment and power

Typical fields of study: Accounting, data processing

Source: Adapted from [210], [222, p. 97] and [245, pp. 38–39, 59–60]

Scales [60] with around 30 types of personality, have been created to provide complementary information for more specific analysis [215].

In choosing field of study, Holland's personality type is, of course, not the single decisive factor. Students also tend to expect strengthening their knowledge in aspects less related to their Holland type [210]. Moreover, there are other classifications of personality types (e.g., the Big Five[7] [229]), which also were found to affect the choice of field of study.

1.2.2 Social Environment: Family, Social Class and Social Networks

One of the most important factors affecting individuals' choices on education is their social environment. It may be called family (or parental) effects, social (or socio-economic) class or social network effects, but the essence remains the same: individuals build their social networks according to the homophily principle and then use them in decision-making. The situation described in the beginning of this section, where the undecided person is likely to choose the field of study chosen by his friend, is a typical example of such influence.

The effects from social networks have been discussed in two perspectives. One of them is *rational action theory* [42, 47]. While, as a typical theory assuming rational behaviour of agents, it is built around individuals making cost–benefit analyses before making their choices, it does not downgrade the effects from social origin. For instance, three mechanisms of social origin effects are built in the model in [47]:

- *Relative risk aversion.* Families seek to ensure that their children end up in the social class at least as high as from which they originate. In other words, they seek to avoid downward social mobility.
- *Primary effects.* Ability (which is related to social background) determines the likelihood of success at different educational tracks. Thus, children have to have a certain minimum ability to qualify for continuing education.
- *Availability of economic resources.* Education is not costless, and certain minimum resources of families (the availability of which differs by social class) are needed for children to continue education.

The relative risk aversion mechanism was empirically confirmed, for instance, in [274].

Another theoretical perspective is *habitus theory* or *social reproduction theory* [43, 44]. It considers two types of individual's capital: cultural capital and economic

[7]The Big Five or the Five Factor Model is a model that defines five main personality traits: openness, conscientiousness, extroversion, agreeableness and neuroticism [72]. This is considered to be the "default model of personality structure" nowadays [184, p. 15].

capital, and two types of elite: cultural elite and economic elite. These elites differ in what capital is available to them and what matters to them. The economic elite uses economic wealth and related lifestyle and values to position itself in the social space and increases the economic capital available to its members. In contrast, the cultural elite seeks to enhance its cultural capital by living a different lifestyle and using different values. Members of different social classes have different cognitive and normative predispositions, referred to as *habitus*, depending on the extent of cultural and economic capital available. It shapes their choices, including those related to education.

Both theoretical explanations predict considerable intergenerational transmission of education level and field of study. Empirical literature finds strong evidence of that. It was found that the choice of field of study at tertiary level depends on parental education level [40, 41] and field of study [281]. Moreover, as proposed by the habitus theory, cultural and economic elites are inter-generationally reproduced by the fields of study of their children [121, 281, 282]. Socio-economic status (SES) has substantial effect on education choices through affecting the perceptions of financial incentives: low-SES individuals perceive higher education as a risky investment and, thus, are less willing to borrow and more attracted by grants or scholarships, more involved in working while studying and choose fields with shorter durations and those perceived as less difficult [288]. There is also evidence that SES interacts with ability in the context of education decisions: higher SES helps low-ability students to enter a more prestigious institution, but high-ability students of low SES choose more practical fields of study, thus, excluding themselves from dominant elites [79]. The social background effects, however, are stable over time: the share of students with more privileged social backgrounds who choose more prestigious vertical (by level) or horizontal (by field of study) education alternatives is not growing [218].

The research cited above studied observable effects from general social environment. When asked directly about using their social networks in their decisions on education, individuals do not always admit that they base decisions on their network, rather than making them themselves. The self-reported influence of non-family members of social network is rather weak: 13.3 % in Romania, compared to 31.2 % of family members [212]; 28 % in Portugal, compared to 80 % of academic reputation [241]. In the Netherlands, it is the weakest of the studied factors [279]. At the same time, members of social networks, especially former students, are used as a valuable source of information about a study programme [241, 279]. Nevertheless, these findings should be used with care, as any self-reported results.

1.2.3 Monetary Costs and Benefits

If one had to name a single theme that runs like a thread through all chapters of a typical book on economics, one would name money. Rational human beings should make the cost–benefit analysis before making any decision and make only those decisions that appear profitable, based on that analysis—at least that is theoretically assumed.

In the context of choosing the education path, it means optimising a function of tuition fees, travelling and/or living costs, the costs of effort to study, the cost of postponing (full-time) participation in the labour market, financial support (such as grants and scholarships) and the expected returns in the labour market after graduation over different combinations of the type of study (e.g., full-time vs. part-time), field of study and education institution. The key question here is to what extent the expected labour-market returns of education affect the decisions of (prospective) students.

It is well known that on average, a higher level of education increases average wage and decreases unemployment risk [198, 206] (but see Sect. 1.3.2), although there is substantial heterogeneity across occupations and fields of study [167, 290]. Prospective students know that and expect an increase in income and better career path after graduating a higher education institution [54, 154, 267].

In the USA, students' perceptions of labour-market outcomes of their education have a large impact on the choice of field of study [143, 194] with the probability of choosing a particular field rising with its future monetary returns [114], in part because students believe that their parents are more likely to approve their choice if it leads to higher income [297]. Studies in other countries also support positive reaction of applicants on monetary labour-market signals [40, 41, 212].

As one would expect, there are limitations on the effects from expected earnings on applicants' choices. Firstly, a *substantial* increase in lifetime earnings is necessary to make individuals choose a field they were not inclined to choose initially [41]. Secondly, in assessing monetary costs and benefits, students take a short-term perspective: study loans and grants are more important in their decision-making than expected future income [288]. Thirdly, individuals already employed before studies are less sensitive to differences in earnings across fields [40]. Overall, studies conclude that students *for whom money is important* tend to choose fields with more stable and solid earnings opportunities [40, 121].

1.3 Individual Behaviour in the Labour Market

I once chatted with a bachelor student in economics. She was happy that she found a place for summer practice. When I asked her how she found it, she said that she'd be supervised by an acquaintance of her father, who had her own firm. I wasn't surprised that she hadn't searched for the job herself, because many find first jobs through friends or relatives. Anyway, I asked her why she preferred this way. Her answer was, "It's easier and faster".

In the labour market, individuals' choices include the following: Should I search for a job through official sources or through my social networks? Should I take a job for which I'm mismatched or continue being unemployed and wait for something better? Should I continue working by my current employer or search for a better job? These topics will now be analysed in turn.

1.3.1 Methods of Job Search and Recruitment

There are two main methods of job search. During *formal* job search, individuals browse databases of vacancies or visit the governmental agency responsible for aiding in job search. During *informal* job search, individuals use their social networks to learn about existing vacancies. The bachelor student mentioned above clearly found the job for summer practice through informal search.

Similarly, firms could also use formal channels by publishing vacancies online or through the mentioned governmental agency or choose to search informally by asking their employees about the candidates they know. Moreover, when choosing among job applicants, firms could ask their employees whether they would recommend some of the applicants if they know them. This behaviour of firms is known as *referral hiring*.

The extent of informal search is substantial for both firms and individuals. Depending on the country, 30–50 % of companies hire by employee referral [34, 133], but the reliance on informal search channels is far more common in small and medium-sized firms than in large firms [133, 162]. Similarly, between 30 % and 50 % of individuals find their new jobs through friends and relatives [113, 193]. Individuals with access to larger social networks use informal job search channels more often [57].

There are two reasons why social networks are so much used in the hiring process [122]. Firstly, they help mitigate the problem of *bilateral asymmetric information*, when both prospective employers and employees do not know the other side's quality. In these settings, they search for more information about one another from personal sources they can trust. Trust plays a crucial role here.

Let's take the job candidate's stand first. Assume you are interested in a particular position in a certain organisation, but you lack some important information about the policies in place there. However, it happens that someone whom you know quite well works there and, most likely, is informed about these policies. If you believe that the organisation, if inquired directly, may either refuse to give you that information or be biased in the qualitative assessment of themselves, you may choose instead to ask the employee you know. Because you trust him, you believe he will provide you with an unbiased view of the situation.

If you put yourself in recruiter's shoes, you'll frequently want to cross-check the candidate's qualifications. If you know someone who knows the candidate—this might be an employee whom you trust or someone from outside your organisation—then you'll consider asking him about the candidate. Or, if you need to fill the open position urgently, you may directly use your network to nominate candidates.

As mentioned above, such behaviour is more likely to be found in smaller organisations, where there is no human resources department with strict procedures guiding the recruitment process.

Empirical evidence generally concludes that referral hiring increases the quality of match between the chosen applicant and the job. This is reflected in higher job finding rate and longer tenure [52], as well as higher initial wages [52] or lower initial wages but higher wage growth [161]. The size of the effect depends on the quality of the referred individual—these effects are stronger at lower levels of skills [52, 161]. The usefulness of the network depends on the interaction between its quality and the skill level of the individual: the job finding rate grows with the number of employed friends for both the high- and low-skilled, but the positive effect from the number of employed friends on wages is observed for the high-skilled only [61].

Secondly, the cost for the firm of searching for a new employee in existing social networks, which are maintained mainly for non-economic reasons, is far lower than when using formal channels. The cost is lowered not only by not needing to prepare job advertisements in various formal channels and pay for their placement, but also by eliminating the time needed to wait for job applications to come and analyse them. If you contact right people, they may promptly give you the contacts of a few candidates. The downside is that the number of potential candidates might be lower and people from your network may not provide accurate enough information about the person they recommend. Nevertheless, the benefits of informal search may easily outweigh these costs.

1.3.2 Overeducation

In which state would you *not* like to end up in the labour market, at least for lengthy time periods? The most frequent answer is "unemployment", which is certainly true. However, I assume, you would also not like to end up in a job where the knowledge and skills you gained while learning for your university degree are completely useless—not because you were suddenly invited to a top-tier position in some industry completely unrelated to the field of study where you graduated, but because you weren't able to find a job where higher education is used at all. This state, with some minor differences in meaning, is known as overeducation, overqualification or overskilling. Roughly speaking, if you are *overeducated*, then your education is higher than what is required in your current job. The reverse corresponding terms are undereducation, underqualification and underskilling. Although the experience of overeducation is considered by employers less harmful than unemployment experience [17], it still has serious negative effects on the individual.

Depending on how mismatch is measured (I'll move to the details on measurement after a few paragraphs), empirical literature shows that from 10 % to 30 % of European labour force are overeducated and around 20 % are undereducated, so that the total mismatch is between 30 % and 50 % [144]. In other words, this is not a marginal problem that could be ignored.

Several theories explain the existence of overeducation through persistent inefficiencies of the labour market (for some reason, ignoring the education market). The *job competition theory* [271] argues that individuals with higher level of education require less training, so they are more likely to be employed. This generates credential inflation, so this theory is sometimes referred to as *credentialism*. The *signalling theory* [248] introduces the costs of education and postulates that, for education to be a valuable signal for firms, its costs should be negatively correlated with employee's productive capabilities. When this correlation is absent, everyone will invest in the given education, making it useless as a signal of high productivity. The *assignment theory* [232, 272] assumes that the actual productivity is the maximum of the individual's own productivity (which depends on qualifications) and job's productivity (which depends on its complexity and technology), so individuals with higher qualifications are assigned to more complex jobs. Then mismatch might arise due to a mistake in the complex assignment process [27], other factors compensating the effect from mismatch on the utility of a position [208] and the macro-level imbalance between the demand for and supply of individuals with different skills [213]. The *job search theory* [251, 252] introduces the costs of job seeking, so that the reservation wage of individuals decreases with the length of job search. Thus, it may fall to the level of the wages in jobs for which they are overeducated. According to the *technological change theory* [90], technological change forces employers to hire workers with better qualifications than their current workers have, but firms cannot change their whole workforce immediately due to high costs and neither are they willing to invest in upgrading the skills of current workforce. Finally, the *theory of differential overeducation* [112] claims that the job search for married couples has two steps: first, the husband finds the best job in the global labour market, and then the wife takes the best job in the local market where the husband found his job. Hence, due to searching in a smaller labour market, the probability of overeducation is higher for wives than for husbands.

Nevertheless, a sizeable part of the scientific community believes that the labour market is actually working perfectly, and mismatch, if it exists at all, is minute and quickly vanishes. If we do observe mismatch, they argue, then this is because we use wrong methods of measuring it, which do not allow us to take into account the *unobserved heterogeneity*—i.e., some unobserved or intangible differences—of workers and jobs. This view is favoured by human capital theory [30] and job mobility theory [238]. The *human capital theory* argues that the firm either provides training to all non-overeducated workers so that they reach the productivity of the overeducated workers or leaves everything as is. In the latter case, because wage is set in accordance with individual marginal product, so that the overeducated are paid below their potential marginal product, they will quit to a more appropriate job. The *job* (or *career*) *mobility theory* concludes that workers agree to spend some time in mismatched positions to accumulate the right amount of skills to receive promotion shortly.

There are indeed several measures of mismatch, which can be classified in normative, statistical, self-assessment and income-ratio measures [247, Table 1]. Shortly, the normative measure is based on a pre-defined mapping between job category and education level, the statistical measure takes average education in the job category and marks those with education, e.g., one standard deviation above the mean as overeducated, the self-assessment measure is self-explanatory and the income-ratio measure depends on comparing actual and potential income. None of these is universally accepted, and each has its advantages and disadvantages.

Overeducation may have certain positive effects: it may boost career movement [124] and upward mobility [88], both in accordance with career mobility theory, as well as increase wage growth inside the organisation [124] and individual productivity [28].

However, because overeducation is persistent over time [18, 37, 95, 115, 156, 227]—i.e., once in that state, a sizeable amount of time is spent there—it was concluded that the "overeducated are penalized early on by an inferior rate of return to schooling, from which they do not recover" [159, p. 183].

Wait a minute. We just read about some theories claiming that overeducation does not exist and, hence, if what we see as overeducation is actually a perfect match between the worker and the job then we would actually *expect* that this situation is persistent over time. If the match is perfect, why change job? There must be some really sizeable negative effects from overeducation that allow to view it as an inferior return to schooling.

So how does that inferior rate of return manifest? Psychologically, through lower job commitment [38], lower job satisfaction [209, 263, 289], more symptoms of depression [45] and decreased participation in extra-role behaviours (i.e., that go beyond what is determined by contract) [3]. You'd actually expect all these effects if the person you analyse perceives their overeducation state as unfortunate.

Materially, through higher earnings compared to the well-matched *in the same job* but lower compared to the well-matched *with the same education level* [130, 228]. For instance, tertiary graduates working in the jobs destined for secondary graduates earn more than the secondary graduates in these jobs but less than the tertiary graduates working in the jobs where higher education is actually used, which seems logical. The only problem is that this is again in contrast with theories claiming that overeducation does not exist. Their "last resort" argument of unobserved heterogeneity was tested extensively in the literature. Panel data methods allowed to decrease wage differences substantially [26, 62, 116, 159, 277, 284], although in some models, the above result was still observed [159], while in others, controlling for heterogeneity did not result in a substantial reduction in the wage effect of overeducation [62, 183]. Another way to control for unobserved heterogeneity is through approximating ability by wages, assuming that higher wage reflects higher unobserved ability. Results are mixed: it was found that the wage penalty for the low-ability overeducated is higher than for the high-ability overeducated [63], that high-ability workers are hurt more [53] and that the effect is pervasive and constant for females but for males exists only in case of low and middle ability [186].

The factors affecting the probability of overeducation can be divided into macro-level factors and micro-level (or individual-level) factors. Much research has been devoted to skills mismatch in relation to macro-level demand factors. Overeducation is found to behave counter-cyclically: the highly educated crowd out the lower-educated during economic downturns [75, 156]. Higher shares of temporary contracts increase overeducation via lowering the selectivity of employers and workers, while higher long-term unemployment decreases it by keeping less-able workers out of labour force [75]. A larger shadow economy does not affect the mismatch likelihood of the locally born, but decreases the chances of overeducation of immigrants by easing the job search process for the lower-skilled; finally, employment protection legislation increases undereducation [9]. Macro-level supply factors also play a role: overeducation on a particular education level increases with the share of population having that education level [93].

Individual characteristics may be more important than macro-level factors [118]. At the microeconomic level, five categories of factors have been found to affect the risk of overeducation: (1) ability, academic performance and personality; (2) gender and age; (3) immigrant background; (4) labour market and job characteristics; and (5) characteristics of education.

Although an individual's ability or academic performance has proven to be difficult to capture in empirical work, research, using various approximations, suggests that graduates with lower ability face a higher risk of overeducation [24, 63, 175, 259, 283]. Personality traits also affect the risk of overeducation and frequently are more important than ability [37, 261].

Empirical evidence about gender effects has been mixed, with approximately equal number of studies concluding that women have a higher skills mismatch risk than men [9, 18, 33, 152, 216, 258, 283, 285] as those finding no difference across sex [29, 37, 63, 64, 115, 116, 253, 295]; a few studies show that men are at a relative disadvantage [92, 156].[8] The literature also disagrees on the effect on overeducation from age. Some studies show that overeducation decreases with age [9, 146, 225, 255] or is U-shaped [259], while others report that age is irrelevant [37, 64, 115, 156, 295]. Overall, these results imply that the young may still have a comparatively higher probability of mismatch after controlling for other relevant factors.

First- and second-generation immigrants face higher risk of mismatch [9, 259], and residence duration seems to have no effect on it [9, 108]. Where overeducation decreases with the length of stay, it was interpreted as immigrants preferring unemployment [253], may be affected by the country's skill-based immigration policy [258], or happens only for specific types of education [31]. Immigrating to a close country or having more knowledge about it dampens the risk of overeducation [9, 258]. Higher-quality education system in the home country increases the chances of undereducation [9], but foreign education—even for the locally born—is not perfectly transferable to the local labour market, as reflected by higher overeducation risk than for the locally educated [253]. The type of mismatch after migration also strongly depends on that before migration [211, 258].

Studies also found that overeducation risk decreases with tenure [29, 115, 146, 152, 295], which allows to conclude that labour market experience acts as a substitute for formal education. Finding a good match in a larger labour market

[8] See also [213] and [92, p. 371, Footnote 17].

should be easier, and in Europe, the risk of overeducation in big cities is indeed substantially smaller than elsewhere [216, 259], although the reverse is observed in the USA [31]. Working without a contract substantially increases the risk of overeducation, and this result is stable across European country groups [259].

Research also shows that graduates in economics, law, and arts and humanities face higher overeducation risk [24, 33, 63, 78, 145, 208, 253]. Studies disagree on the sign of the effect from educational attainment (in terms of years or level). While some find that the risk decreases with attainment [24, 29, 146], others report that higher years of education [108, 145] or a university degree [115, 295] increases the risk of overeducation.

1.3.3 Quitting the Job

By now, I have reviewed the ways how individuals search for a new job and how they may end up in a job for which they have inappropriate education. To close the story, I'll now review how individuals decide to quit the current job.[9] As in job search, social networks play an important role in this process.

According to Fig. 1.4, individuals first form attitudes towards their job depending on three large groups of factors: their own characteristics and those of the job and the organisation. These attitudes, after passing through cognition moderators, determine whether withdrawal cognitions (most importantly, intention to quit) and/or withdrawal behaviours (e.g., lateness, absenteeism, on-the-job search) develop. Mere intention to quit does not necessarily lead to quitting—action moderators determine whether the intention materialises in voluntary turnover.

Two attitudes are typically found to be strong predictors of turnover: job satisfaction[10] and organisational commitment.

A classic definition of *job satisfaction* is "pleasurable emotional state resulting from the appraisal of one's job as achieving or facilitating one's job values" [177, p. 317]. It was later reformulated from "emotional state" to "affective orientation" towards the employee's work role [151] or "affective reaction" to the job resulting from the comparison of the actual and desired work-related outcomes [74] or, in contrast to the previous two, an "evaluative judgement" (unrelated to emotions or

---

[9]But see, e.g., [131, 205] on the antecedents of *collective* (group, unit or organisation-level) turnover.

[10]Another intensively studied effect from job satisfaction is on job performance. On average, job satisfaction and job performance have a correlation of 0.30 [150], although there are several moderators of this relationship. For instance, the correlation is higher (0.52) in high complexity jobs, in part because these jobs have fewer constraints on employee behaviour [150]. It is also higher for individuals with higher affective–cognitive consistency, i.e., when the emotional feeling about something (the affective component of the attitude) is consistent with the beliefs or thoughts about it (its cognitive component) [234, 287]. Performance in itself affects turnover, as will be discussed below.

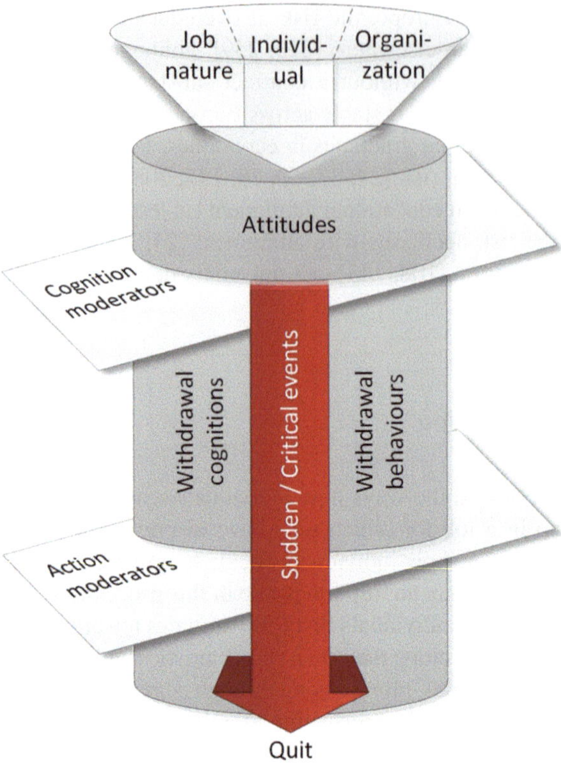

Fig. 1.4 Conceptual summary of job quit processes. *Source:* developed based on [123, 142, 249, 256]. Individuals form attitudes towards their job depending on job nature, their own characteristics and those of their organisation. Without any sudden or critical events, which lead to quitting immediately, cognition moderators determine whether and to what extent withdrawal cognitions and/or withdrawal behaviours take place. Action moderators regulate whether withdrawal cognitions and behaviours lead to actually quitting the job

feelings, so *not* an affect) about the job [292]. General-population studies show that job satisfaction is primarily a function of job content (e.g., challenge level of tasks, education or skill mismatch), job-related risks (e.g., health risk, job security), monetary compensation, working conditions (e.g., opportunity to pursue own ideas, to manage working time) and social environment [195, 244, 263, 286].

Whether through attitudes or not, the following groups of factors were found to influence the intention to quit. Firstly, it is the imbalance between job demands and resources provided on the job, which was captured by several theoretical frameworks. The *conservation of resources* model [137] argues that people aim at retaining, protecting and building resources, while a possibility to lose them causes stress. The *job demands–resources* model [19, 89] adds demands on the scene and argues that if the available resources are insufficient to meet job demands, stress occurs. Similarly, the *effort–reward imbalance* model [239] explains stress by an imbalance between the efforts spent at work and occupational rewards received in return. The *job demand–control–support* model [148, 153] divides resources into job control (by the individual) and work-related social support. The results of the imbalance between demands (efforts) and resources such as burnout [12, 132], stress [13, 110, 132, 169], working overtime [250], inability to delegate [132] and dissatisfaction with rewards [91, 250], but also role ambiguity [12, 13, 123, 132], all increase the intention to quit and, ultimately, lead to quitting.

Social ties play a dual role in the turnover process. On the one hand, in accordance with the mentioned theories, social ties with co-workers and management play a supporting role to help the individual cope with challenging job demands. By reducing (a possibility of) stress, social support increases job satisfaction, which further translates into lower quit intentions [10, 13, 51, 71, 73, 98, 132, 197, 207, 226]. On the other hand, the role of social ties is substantially larger than that, which is pinpointed by the term *job embeddedness*—"the extent of an employee's 'stuckness', or enmeshing, within a larger social system" [293, p. 201]. Job embeddedness has three dimensions: *links*—"formal or informal connections between a person and institutions or other people", *fit*—"an employee's perceived compatibility or comfort with an organisation and with [their] environment" [189, p. 1104] and *sacrifice*—"the perceived cost of material or psychological benefits that may be forfeited by leaving a job" [189, p. 1105]. Generally, job embeddedness is a strong motivator to stay in the job [76, 107, 189, 293].[11] It can also be divided into internal (on-the-job) and external (off-the-job) embeddedness; both have turnover-decreasing effect[12] [147], although there is some evidence that external networks might instead foster the intention to quit [197, 223], for instance, because employees learn about new outside opportunities through them.

From individual-specific factors, being married [4, 123, 169, 237, 246] and having children [123, 132, 235] decrease the intention to quit; it also decreases with age [76, 104, 123, 132, 237], tenure [107, 119, 123, 132, 160, 197] and performance (especially, in the presence of performance-contingent rewards) [123].[13] Overeducation [224] and overskilling [11, 149, 187], on the contrary, increase the intention to quit and actual turnover.

Besides demands and resources, other job-/organisation-specific factors that increase the intention to quit include working in private (compared to the public) sector [169, 246], physical hazards and work in night shift [73], lack of promotion possibilities [94, 104, 123, 132], lack of training[14] [123, 149, 160] and downsizing rate in the organisation [276].

External factors also play a role in the voluntary turnover process. As one would expect, the intention to quit shows up faster and materialises into actually quitting the organisation with a higher chance if the unemployment rate is lower [4] and there

[11]Co-workers' job embeddedness acts in the same direction [107].

[12]A similar result is reported in [66], where internal and external network sizes were used instead.

[13]Some studies point at a U-shaped effect from the number of children [4, 237] and age [235, 246].

[14]Training was also found to increase the intention to quit [94, 160].

are more attractive alternative jobs [76, 123] or other pull factors are present [236]. Another important set of moderators are various personality traits [300], human resource management practices used by the organisation such as performance-contingent rewards [104, 123, 165] or high-involvement work practices [73] and job embeddedness [293].

Chapter 2
Complex Adaptive Systems and Agent-Based Modelling

> *A hearing of the House of Representatives Committee of Science and Technology [...] aimed [at questioning] the wisdom of relying for national economic policy on a single, specific model when alternatives are available. [...] One of the most promising options was the topic of a workshop [...] funded by America's National Science Foundation. [...] They were to explore the potential of agent-based models of the economy.*
>
> *The Economist [269]*

In a labour–education market system, there are many individuals and firms with adaptive behaviour. As we have seen in the previous chapter, networks are prevalent in LEMS and play an important role in many decisions of its actors. Thus, LEMS can be analysed as a complex adaptive system (CAS). Agent-based modelling (ABM) is typically used for such purposes, and the next chapter will dig into details of various ways of applying ABM in modelling LEMS. To be ready for it, we first have to understand the motivation behind and the details of this method. This is what will be discussed here.

This chapter starts with a short description of the nature of ABM. Then it discusses why this method should be applied to LEMS, compared to standard economic modelling methods and other simulation methods. The chapter continues with a detailed review of the process of ABM. Finally, its limitations are noted.

2.1 What Is Agent-Based Modelling?

The pioneers in using ABM to study social phenomena were James Sakoda and Thomas Schelling (for their early research, dating back to the 1970s, see [230] and [233], respectively). In the 1980s, after analysing biological and social models, research groups at the Massachusetts Institute of Technology started studying the possibilities of using multi-agent systems in problem solving.

A. Tarvid, *Agent-Based Modelling of Social Networks in Labour–Education Market System*, SpringerBriefs in Complexity, DOI 10.1007/978-3-319-26539-1_2

Typically, intelligent agents are software problem solvers with four characteristics [179]:

- *Situated*—react to the signals from their environment, their feedback can change the environment
- *Autonomous*—control their actions and internal state and are able to act without direct manipulation by other agents
- *Flexible*—not only respond to environmental signals, but also can plan their actions in response to possible future signals to reach their goal
- *Social*—can interact with other agents

An *agent* is, thus, defined as "an element of a society that can perceive (often limited) aspects of its environment and affect that environment either directly or through cooperation with other agents" [179, p. 19].

In applications of ABM to economics and finance, these features remain unchanged. Without any global controller, autonomous agents representing various model entities such as individuals, organisations or governments—depending on the purpose of the model—are making choices, performing actions and communicating with each other in the aim of achieving some specific goal(s).

In artificial intelligence, ABM was initially used as an alternative to approaches that use mathematical logic with assumptions like perfect rationality of system components to specify component behaviour. Instead, the goal of ABM is to observe and then interpret the dynamically evolving structures during simulation runs [86]. Thus, computer simulations are run, and simulation output is then interpreted by the modeller in the context of model purpose.

One of the unquestionable benefits of applying ABM to study the dynamics of some system is that one is not constrained to specifying system behaviour in the form that is mathematically soluble. Firstly, this allows building complex behaviour models using many if–then rules. This is important, because behavioural patterns are captured by flow charts much easier and more naturally than by systems of equations. Secondly, simulations can contain many random elements governed by different probability distributions. Applications in economics and finance, as opposed to some other fields ABM is applied in, nearly always require randomness in the specifications of agent or system behaviour. Mathematical modelling in such settings quickly becomes prohibitively difficult.

As in any model, there is a trade-off between the correspondence with reality and complexity in ABM. Each agent-based model depends on a set of parameters. Each parameter can be defined by a single value or a set of values or drawn from a probability distribution. If there is no randomness in the model, it is sufficient to run it once for each combination of parameters. Otherwise, for each combination of non-random parameters, the model should be run several times, for obvious reasons (of course, more is better). Because, as it was already mentioned, in agent-based models in economics and finance, randomness is always included, one is forced to define parameters as precisely as possible to minimise the number of model runs. Moreover, if many parameters are not defined precisely using empirical data or theory, the simulation gives less reliable results. The reason is that the modeller has to run the simulation many times, and he does not really know, which parameter combination corresponds to the real world; thus, only results that appear consistent in all runs might be treated as reliable [49]. Nevertheless, it is always recommended to check the sensitivity of the model, i.e., the stability of its output with respect to small changes in parameter values.

A more detailed discussion on how ABM is done is postponed until the end of the chapter. Now, I'll explain why we need ABM in modelling LEMS at all.

2.2 Why to Use Agent-Based Modelling...

ABM is a natural method of modelling LEMS. Why? Because ABM is a natural method of modelling CASs, an economy is a CAS and LEMS is a part of the economy. There are now three questions: What is a complex adaptive system? Why is ABM preferable to the modelling approaches that have been traditionally used to model economic systems? When should we use ABM and not other simulation methods? These questions will now be discussed in turn.

2.2.1 ...for Modelling Complex Adaptive Systems?

The concept of a *CAS* has been discussed in many different fields in the last three decades, such as social–ecological systems [172], management [6, 102, 278] and education [176, 182, 185]. To understand, why, look on some of the definitions of a CAS given in Table 2.1. For sure, you can immediately give a few examples of systems that conform to these definitions—e.g., stock market, immune system, the Internet. The difficulty (or, in other words, complexity) of studying CASs is that conventional (especially, conventional mathematical) modelling methods have limited usefulness in that process. Earlier I claimed that we might consider LEMS as a CAS, and now I will illustrate the definitions from Table 2.1 using the key elements of LEMS.

The primary—in some sense, atomic—element of LEMS is an individual. In the definitions in Table 2.1, these are called components, actors or agents. I'll stick to the latter term. Typically, the number of agents in a system is considerable but finite. In an economy, there are many agents, from a few tens of thousands to over a billion (depending on country size). Most of these agents participate in LEMS—generally, on its labour market side (others are simply too young, too old, have other values in life or are victims of circumstances).

An agent itself is a collection of properties and behaviours. A fundamental characteristic of agents is the heterogeneity of their properties within a given CAS. For instance, individuals are typically of different age, they have different education, experience and abilities, which makes a particular worker more or less suitable for performing particular tasks. Agent's properties do not have to be fixed over

Table 2.1 Definitions of a complex adaptive system

A system that involves "many components that adapt or learn as they interact" [139, p. 1]
A system whose macro-level properties "emerge from interactions among components and may feed back to influence the subsequent development of those interactions" [171, p. 431]
A system consisting of a network of interacting adaptive agents that exhibits a dynamic aggregate behaviour that emerges from the individual activities of the agents but that can be described without a detailed knowledge of the behaviour of the individual agents; agents are adaptive in the sense that their actions in the environment can be assigned a value that, due to agents' behaviour, increases over time [140]
An open system with "multiple components, which interact with the environment," connected into "self-organising networks," through which both the components influence the system and the system influences each component, that is characterised by co-evolution and adaptation and generates emergent behaviours [101, p. 486]
A system of autonomous cooperating agents that has the properties of self-similarity, emergence, self-organisation, distributed control and continual adaptation [278]
A system "composed of many interacting, intelligent and independent actors", whose behaviour is "an outcome of physical, psychological or social rules", with "the capacity of adapting to changed circumstances" resulting from the interactions among agents, the interactions of the system with the environment and the system's history [185, p. 132]

time: while individual's ability level might be considered innate, their age and experience—but also education level—(may) change during their life.

Adaptation is required at system level—that's why these systems are called complex *adaptive* systems—but *not* at agent level. Agents can adapt, in the sense of actually changing their behavioural rules in response to changing environment or the feedback of the environment or other agents on their actions. Typically, that means that they constantly evaluate the gains from previous behaviour and seek ways to increase them in the future. But they can also continue following the same rules if these were defined broadly enough.

At the same time, nowhere is it claimed that agents *optimise* their utility, fitness or some other measure of gains. They can actually decide *not* to increase the gains if they are satisfied enough with what they have.[1] As an example, consider an unemployed individual seeking a job. He has two channels to get the information about jobs: formal and informal, as described in Chap. 1. We may assume that the individual will use the formal channel more frequently if it proves to be more effective than the informal channel. If the informal channel suddenly starts to perform better, individuals may start using it more frequently. Certainly, it looks as adaptation, but does it come from a change in fundamental rules of behaviour or is it simply the result of the same rule (use more frequently whatever works better) applied in a different environment? While it seems to be a philosophical question, it actually influences the actual implementation of the model. Anyway, it is unreasonable to assume that, given historical experience, the individual finds the *optimal* allocation of resources between formal and informal search.

[1]Such behaviour is an example of *bounded rational* behaviour. It is called *satisficing* behaviour, from the words "satisfy" and "suffice" [243].

Another element crucial for a CAS is network. Indeed, much of the non-linear behaviour of system output is typically caused by networks connecting agents, through which they exchange information and/or goods and services. Networks normally can evolve over time, with some edges added and some deleted (vertices are naturally added and deleted, as agents enter and leave the system). We already saw how networks of friendship, acquaintance and relationships are created and used in LEMS in Chap. 1, and it is easy to imagine how these can (though not necessarily will) distort the results of a theoretical model that does not include networks.

The final property of a CAS is *emergence*, the phenomenon when "the interaction of autonomous or quasi-autonomous components of the system [...] leads to higher level functionality that is not present in any of the individual components" [16, p. 111]. In other words, agents in a CAS are able to create macro-level structures and dynamics that could not be predicted using only knowledge about micro-level characteristics and behaviour of agents. Again, networks play a very important role in shaping the emergent patterns of system behaviour.

The definition of an agent given in the beginning of this chapter (Sect. 2.1) clearly corresponds to what is understood by an agent in a CAS. Furthermore, object-oriented programming (OOP), which is typically used in ABM, is also a natural paradigm for programming CAS agents with a set of properties and behaviours (*fields* and *methods*, respectively, in OOP parlance). Finally, ABM naturally includes the possibility to model networks connecting agents. That's why ABM is a natural approach to modelling CASs.

2.2.2 *...If Traditional Economic Modelling Exists?*

For historical reasons, economics has used analytical modelling based on the physics of the nineteenth century. Consequently, traditional models sacrifice their ties with reality in favour of being mathematically soluble. There is a vast critical literature on neoclassical economics, reviewing which in detail is beyond the scope of this book. If you're interested, see [32] for an excellent review of historical peculiarities that were crucial to the development of neoclassical economics. If you'd like to read a more rigorous treatment of the problems resulting from the typical assumptions of traditional models, see [138, 157, 158, 166], among others.

The main flaws of traditional economic modelling are summarised in Table 2.2. These can be grouped into five categories: (1) perfect rationality of individuals, (2) homogeneity of agents in models, (3) lack of interaction between individual agents, (4) analysis of equilibria without bothering about whether, when and how these will be reached, and (5) other simplistic assumptions.

The table also notes how ABM helps to overcome these flaws. This is possible exactly because this method of modelling is extremely flexible. The modeller is not constrained with specifying the model as a system of equations that must be soluble, which allows to create the model as complex as he or she wishes.

Table 2.2 Traditional economic modelling vs. agent-based modelling

Traditional economic modelling	Agent-based modelling
Rationality: • Humans can figure out the future of the world, at least on average • They act to maximise their welfare Problems: • Humans have bounded capacity to figure out how to behave • Finding Walrasian/Nash equilibria is too computationally difficult for an ordinary human • The assumed rational behaviour lacks procedural/algorithmic bases • Human behaviour systematically departs from rationality requirements	Bounded rationality: • Agents inspect only local environment for possible utility gains • They take actions that they believe to lead to satisfactory, not necessarily optimal, outcomes, at least with high probability • They have little global information, and the information they have may be significantly out-of-date • Acquiring global information may be possible but costly • They have limited knowledge of their own preferences and understand them after trying different alternatives
Agent homogeneity: • Agents are homogeneous or with very limited heterogeneity • Allows to analyse the representative agent Problems: • No plausible formal justification for the assumption that the aggregate of individuals acts itself like an individual maximiser • Representative agent's reaction to policy changes need not reflect those of individuals, and its preferences may be opposed to those of society • To comply with empirical tests showing complicated dynamics, the representative agent should possess very unnatural characteristics	Agent heterogeneity: • Each agent is modelled individually: there is no representative agent • Agents have access to different resources and information • The behaviour of each agent is different from the average behaviour • Each agent acts in its unique environment

(Continued on next page.)

Because the model is created as a computer programme, the behaviour of agents could be easily described by a set of if–then rules, in effect, putting a flowchart describing agent behaviour into computer code. This is beneficial, as behavioural patterns are clearly more easily and naturally captured by flowcharts than by systems of equations, as already mentioned above.

Simulations can also contain many random elements governed by different probability distributions. Applications in economics nearly always require randomness in agent behaviour, and this behaviour may not always fall under analytically well-studied distributions such as normal distribution. Mathematical modelling in such settings quickly becomes prohibitively complicated.

Table 2.2 Traditional economic modelling vs. agent-based modelling (cont.)

Traditional economic modelling	Agent-based modelling
Lack of interaction: • Agents interact indirectly, through aggregate economic variables like price vector or unemployment level • Agents do not set the levels of these aggregate variables Problems: • This does not square with the way real-world individuals interact	Interaction through networks: • Permits truly direct non-anonymous interactions • Makes information local, hence agents (1) have locally purposive behaviour than can be either reinforced or not by the global environment and (2) cannot optimise over all states of the world • Allows us to use our knowledge on human behaviour, motivation and relationships in building these models
Agent-level equilibrium: • Only individual-level fixed-point equilibria are of interest • Typically, any fluctuations in the economy are assumed to be exogenous Problems: • Such systems are *thoroughly static* in the von Neumann–Morgenstern sense: no agent has any incentive to unilaterally change • Thus, there is no mechanism for endogenously creating novelty • The source of economic change is sought in non-economic phenomena	Emergence: • Macro patterns emerge from micro-level behaviour and interactions • This opens the way for the evolutionary process to create novelty in the system • The systems under consideration are open and dynamic, not closed and static
Simplistic assumptions: • Many simplifying assumptions introduced for analytical tractability • Typical assumptions: linearity, homogeneity, normality, stationarity Problems: • Many of them fail empirical tests • Making wrong assumptions leads to solving the wrong problem	Complexity: • The use of simulations allows to add more real-world complexity into the model without concerning the detrimental effects on analytical tractability

Source: Compiled from [14, 16, 22, 32, 86, 158]

2.2.3 ...*If Other Simulation Methods Exist?*

Some of the benefits of ABM over mathematical modelling are common for all simulation methods. These can be divided into individual-level methods, where the two commonly used methods besides ABM are *cellular automata* and *microsimulation*

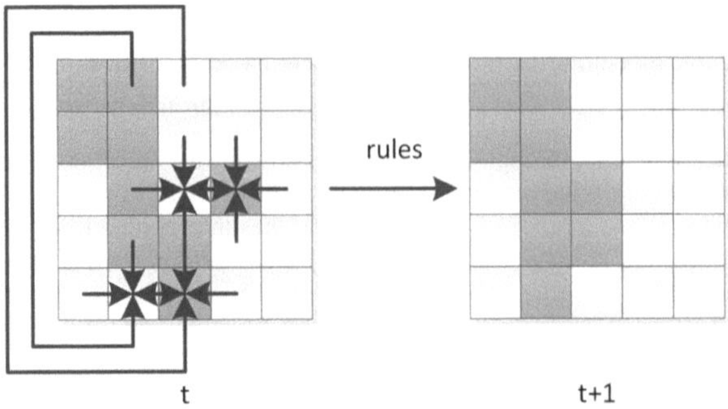

Fig. 2.1 Illustrative example of a cellular automaton. Each cell has a binary value (*white* or *grey*). The value of the cell at time $t + 1$ depends on the values of the cell and its von Neumann neighbourhood (its neighbours above, below, to the left and to the right) at time t. The 5×5 grid is assumed to be a torus (note how that affects the values of the border cells). The rule is as follows: if most (i.e., 3 or 4) of the neighbouring cells are of the same colour, the cell changes colour to match that colour. Otherwise, the cell does not change colour. *Arrows* show why four cells change colours

models, and macro-level methods, such as *system dynamics*. Why then should you choose ABM over its competitors from the simulation toolbox?

Cellular automata are frequently used to study complex systems. This modelling method considers a grid[2] and places the modelled entities in its cells (hence the name of the method). In a typical cellular automaton, entities have a single attribute, whose values depend deterministically on the values of this attribute of the entities located in the neighbouring cells, see Fig. 2.1 for an example. This characteristic makes cellular automata an important method for studying spatial phenomena [25, 231, 275], but also social dynamics such as segregation [134]. The dependence of attributes on those of neighbouring cells can, in principle, be considered as the effects from network connections, with two limitations. Firstly, the dependence is on *neighbouring* cells, so that only *local* interactions affect the state of the entity. Secondly, the neighbourhood that affects the entity is typically fixed, meaning that the network is also fixed. This limits the applicability of cellular automata to more elaborate phenomena—such as individual behaviour in LEMS or where networks have to change over time. That's why cellular automata, where they were used in modelling economic phenomena, were typically only a part of the model. For instance, the famous Sugarscape model [103] combines ABM with cellular automata, as agents there move around a 50×50 lattice, where each cell contains sugar that agents are able to consume. The cellular automaton there only determines the grid and how sugar supplies grow there, while the movement of agents along the grid and their consumption behaviour are controlled by the agent-based part of the model.

[2]The opposite borders of the grid are sometimes "glued" to make a torus. Two-dimensional grids are the most frequently used in applied research, but one-dimensional (e.g., a ring) automata or automata with $n > 2$ dimensions were also analysed.

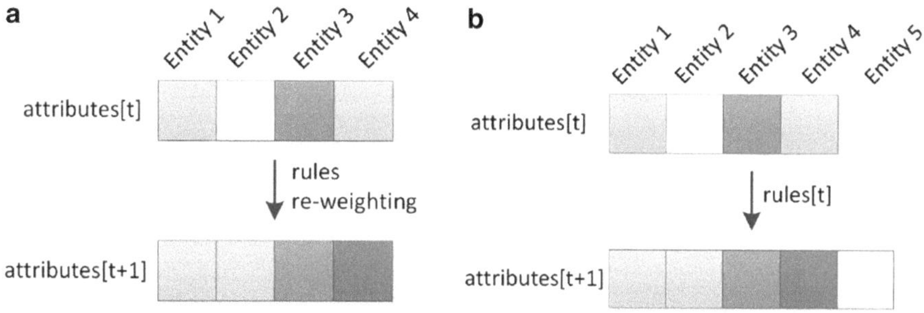

Fig. 2.2 (**a**) Visualisation of static microsimulation models. Attributes are changed by static rules and re-weighting the static population according to external information. Focus on short-term. (**b**) Visualisation of dynamic microsimulation models. Attributes are changed by dynamic rules. The population can be changed (e.g., birth/death is modelled). Focus on long-term

Pure *microsimulation models* contain entities having several attributes that change over time following deterministic or stochastic rules. These models can be divided into static and dynamic models, as illustrated in Fig. 2.2a, b. In the former, a static population is generated based on a representative survey and static rules and re-weighting are applied to it to simulate changes over time. In the latter, the population is actually changed, simulating demographic processes such as birth, death and marriage. Microsimulation has been mainly used to analyse the effects from policy changes (see [36, 174] for examples of actual models and [109, 174] on their construction and use in policy analysis). The two classes have different focus: static models are typically focused on short-term policy change effects, while dynamic models are used to analyse long-term effects. While useful for policy analysis, microsimulation models have the following drawbacks [36]:

- They should be built on large representative high-quality datasets, as model entities represent respondents from the dataset
- They are computationally expensive (although this issue has become less important due to a proliferation of relatively cheap computing power)
- They only model one-way effects from the policy on individuals, ignoring feedback
- They are less strong in behavioural modelling (in particular, they ignore social networks and do not model evolutionary changes in behaviour)
- They are difficult to validate

To sum up, in comparison with cellular automata and microsimulations, ABM allows for (1) a more detailed modelling of behaviour, adding rich interaction effects at the micro-level through networks, and (2) feedback from the micro- to the macro-level. Not surprisingly, a trend towards uniting all three individual-level simulation approaches exists, combining the strengths of each; see, e.g., [36] for a discussion if you're interested.

A macro-level alternative to ABM is *system dynamics*. It models the system as consisting of stock and flow variables, the relationships among them and time

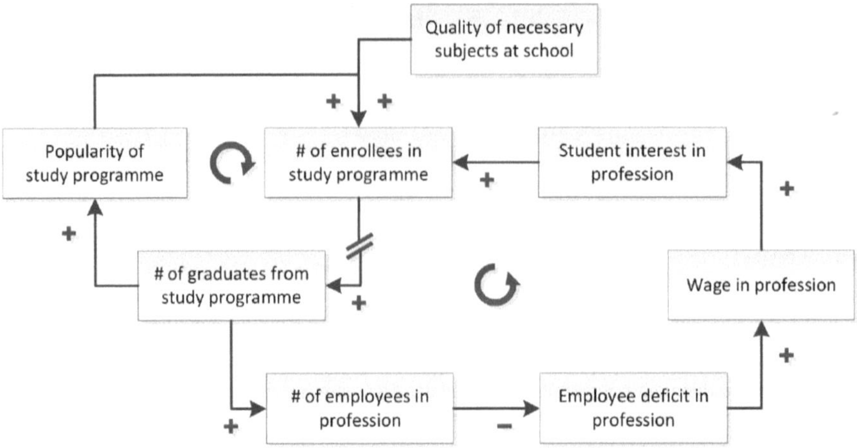

Fig. 2.3 Example of a causal loop diagram. There are two feedback loops in the diagram. The *top left* loop is a reinforcing feedback loop, whereby the number of enrollees in a study programme increases the number of graduates from (with a delay) and, hence, the popularity of the programme, leading to more students enrolling in the programme. The *bottom right* loop is a balancing feedback loop, whereby higher employee deficit in a profession increases the interest of students in it through higher wages, causing them to enrol in and graduate from the respective study programme, which ultimately decreases the deficit of employees in the profession. The quality of subjects taught at secondary school level that are necessary for the study programme is an external variable; it positively contributes to the number of enrolled students

delays. *Stock variables* represent aspects of the system that accumulate (or deplete) over time, while *flow variables* represent changes in stock variables. Relationships in system dynamics models are represented by *causal loop diagrams*, which are based on positive (or reinforcement) feedback loops and negative (or balancing) feedback loops. In a *positive feedback loop*, an increase in one variable ultimately, through the loop, leads to a further increase in it; in a *negative feedback loop*, it leads to a decrease in it. The incorporation of feedback loops and delays in the propagation of information through the loops is what makes system dynamics models able to capture the highly non-linear dynamics of complex systems. Then external variables, which do not participate in any feedback loop, are added and connected to the relevant internal variables. The exact change in a variable resulting from the change in the variable it is connected to is given by a formula, which is derived through precise knowledge or assumptions about how the system's elements function. See [155, 254] for examples of system dynamics models of education market and Fig. 2.3 for an example of a causal loop diagram in the context of LEMS. System dynamics has difficulties with modelling strong spatial or geographical components, dynamically interacting networks of agents, discrete decision variables and constraints on decision variables [204]. ABM, on the contrary, is able to successfully incorporate these aspects.

2.3 How to Do ABM?

The typical process of building an agent-based model follows these steps:

1. Identify the goal of the model
2. Create the abstract structure of the model
3. Implement the model (e.g., using a specialised software package[3])
4. Set the values for model parameters
5. Validate the model
6. Run the model
7. Analyse the output of the model
8. Do sensitivity analysis

2.3.1 From Goal to Implementation

Every model has its purpose reflected in the research question. Four general types of research questions that may be pursued by using ABM are distinguished in [49]. Firstly, the aim might be to describe the characteristics of the modelled system and analyse the relationships between them. Secondly, the model might be built as a prediction tool. Thirdly, the modeller might aim at testing hypotheses about causal relationships, in which case the hypotheses are translated into model assumptions, which are then checked against empirically observable behaviour of the modelled system. Finally, the goal might be to categorise the modelled system into sub-classes based on its behaviour in different specifications.

Subject to the chosen research question, an abstract (mathematical) model is specified. Yes, in ABM we still use mathematics to define various equations determining the dynamics of the characteristics of agents or the environment. In contrast to a pure mathematical model, we don't write *all* dynamic elements as equations and don't try to solve the resulting system or study its dynamics purely analytically. The abstract model is then implemented in a general-purpose or specialised software package, depending on the modeller's programming skills and model complexity.

[3]Several packages are available on the market. Some of the most popular include NetLogo, Repast, MASON and Swarm. Reviewing the available packages is outside the scope of this book. If you're interested, see, e.g., [214, 298] for a comparison of the most popular platforms and [203] for a more comprehensive review.

2.3.2 Parametrisation and Calibration: Setting Parameters

Even if the general specification of agent behaviour corresponds to reality, the modeller should pay considerable attention to setting model parameters. To make a bit extreme example, if an LEMS model is created for a specific country—say, France—then it should reflect the particularities of LEMS in France and not in the USA.

This is actually one of the longest steps in the model building process. There are two reasons behind that. Firstly, not all parameters of the model are straightly available in macro- or micro-level statistical data and might have to be estimated or approximated. This will be discussed at length in Chap. 3. Secondly, in many cases, there will be parameters for which there are in principle no data available, and these will have to be set to the values at which the model generates the expected output. Searching for these "right" values might take a long time. Generally, the model should be run for every possible value of every such parameter. In practice, the model is run for only several of these values for a given parameter. An implicit assumption behind this action is that model output changes sufficiently smoothly with these parameters and, thus, we don't miss sudden exceptionally high or low output values. The number of runs to be executed increases rapidly with the number of such unknown parameters, which is why this number is typically kept very small.

In the following text, I will call the process of setting the values of parameters using available data (either directly or after estimation or approximation) as *parametrisation*. The process of setting the values of parameters for which there are no data (described above) will be referred to as *calibration*. Defined this way, calibration takes substantially more time than parametrisation, hence the need to disentangle the two. At the same time, I'll refer to both actions as *model calibration*. So when I mention calibration in the context of a *single given parameter*, I mean that I have no data available for it and have to find the value(s) at which the system gives the expected output. When I mention calibration in the context of the *whole model*, I mean performing both parametrisation and calibration, as needed, for its parameters.

There are three main approaches to calibration: indirect calibration, Werker–Brenner approach and history-friendly approach.[4] *Indirect calibration* works by setting parameter values so that the model generates a set of *stylised facts*, the well-known facts that have been repeatedly proven empirically. The *Werker–Brenner approach* advocates the use of Bayesian inference procedures and expert opinions in choosing between competing models. Finally, the *history-friendly approach* aims at finding parameter values that replicate historical dynamics. Each approach has its drawbacks; these are described in Table 2.3.

[4]While in [106], these are called validation approaches, their aim is "to reduce the number of model parameters and the space of possible 'worlds' that are explored by tying the model down to an observed empirical reality" [106, p. 206], which is closer to setting parameters (calibration) than to checking the model for correctness (validation).

Table 2.3 Comparison of approaches to model calibration

Indirect calibration:

1. Identify a set of stylised facts to be reproduced/explained by the model
2. Build the model, keeping micro behaviour close to empirical and experimental evidence
3. Use empirical evidence on stylised facts to restrict parameter space
4. Explore the causal mechanisms that underlie stylised facts or explore the emergence of fresh stylised facts

Problems:

- Micro and macro parameters not calibrated using their empirical counterparts
- Unclear how to interpret alternative parameter values in the sub-region of the parameter space that appears after step 3

Werker–Brenner approach:

1. Use empirical knowledge to calibrate initial conditions and the ranges of parameters
2. Perform empirical validation of the outputs for each of the model specifications derived from step 1 (using Bayesian inference procedures)
3. Ask expert opinion

Problems:

- Assessing fitness among a class of models does not automatically help to identify a true underlying model
- Calibration tends to influence the models developed
- The quality of available empirical data might be poor
- Unclear whether the data generating process is ergodic[a]
- Unclear what should be the initial conditions
- The observed parameters might be time-dependent
- Unclear to what extent predictions take into account data outside the current regime

History-friendly approach:

1. Build models on various detailed data—from detailed empirical studies to anecdotal evidence to histories written about the industry under study
2. Compare model output to the "actual" history
3. Having identified "history-replicating" parameters, seek for "history-divergent" results

Problems:

- In practice, such modelling is based on the history of a few key players rather than of the entire industry
- Impossible to get all relevant data to build an empirically sound model
- Limited attention given to sensitivity analysis
- An individual simulated trace that resembles the actual history may or may not be typical of the model
- If several combinations of parameters produce an identical output trace, it's unclear which combination is correct for initial settings
- Difficult to construct counter-factual histories

Source: compiled from [106]

[a] Informally, if the underlying stochastic process is *ergodic*, we can take its single sufficiently long realisation and, by studying it, infer about all possible realisations of this process. If it is not, the generalisation of the properties of a single realisation of the process to all its realisations is impossible

2.3.3 Validation: Checking the Model

Validation[5] is the process of checking that the model is a correct representation of reality or, in other words, that the output of the simulation is comparable to that of the real system [86].

In the context of ABM, it is particularly important to validate model structure and model behaviour [221].[6] *Model structure validation* includes checking that the model structure is consistent with the relevant descriptive knowledge of the system (*structure validation*)[7], model behaviour makes sense even when parameters take on extreme values (*extreme condition validation*) and the important concepts for addressing the problem are endogenous (*boundary adequacy validation*).

Model behaviour validation includes checking whether the model generates the symptoms of the problem under study, behaviour modes, phasing, frequencies and other characteristics of behaviour of the real system (*behaviour reproduction*); whether anomalous behaviour arises if an assumption of the model is deleted (*behaviour anomaly*); whether the model can reproduce the behaviour of other examples of systems in the same class as the model (*family reproduction*); and whether the model behaves properly when subjected to extreme policies (*extreme policy validation*). Besides reproducing historical behaviour, some researchers argue for using *predictive* (as opposed to *descriptive*) validation, i.e., checking the ability of the model to match future, yet unknown, behaviour to make "exhaustive analysis of a model meant to reproduce reality" [35, p. 247].

2.3.4 Running the Model

After the model has been implemented, its parameters set and structure and behaviour validated, it is ready to run. If there is absolutely no randomness in the model, it is sufficient to run it once and study its output. But typically, there is at least one random parameter—e.g., a random change in the sales of a specific firm or selection of a random agent with whom to communicate from the list of agents. In this case, it is necessary to run the model multiple times to see how it performs with different sequences of random numbers.

---

[5]Frequently, validation is mentioned together with *verification*, which means checking that the computer programme executes correctly, i.e., the system implemented by the programme corresponds to the conceptual model the modeller intended to study. Informally, verification means checking that "you are building the thing right". Verification is an important step, but it should be performed at the implementation stage.

[6]In [221], authors also consider validating the implementation, which is the same as verification.

[7]*Input* or *ex ante validation* [35] is also concerned with this subtype of model structure validation, but the authors also include validating parameter values through analysing empirical data, for which there is no need if calibration was properly performed.

The question is, of course, how many times to run the model. There are two possibilities. One is to select a sufficient from the statistical point of view number of runs—e.g., 30 or 50 or 100—depending on how much resources one run of the model uses, and execute these runs. The second option is to choose a set of key output variables and run the model until their distributions stabilise [178]. This option might be more attractive than the first one: there might be no reason to run the model for 500 times if output stabilises already after 70 runs. No one guarantees that, though, and you could easily have to run the model more times than you expected to do. Of course, following this approach requires additional effort from you (check how the distributions of the key variables change with the number of runs and execute additional runs if necessary), but the result of that is certainty that running the model more times will not result in substantial changes.

2.3.5 Analysing Output and Its Sensitivity

Methods used in analysing output depend on the research question and range from the description of the dynamics of some characteristic of the model to heavy statistical analysis.

An additional analysis required after analysing output at basic assumptions and parameter values is *sensitivity analysis*. It amounts to checking how the output of the model changes if its assumptions or parameters are altered. Sensitivity analysis has two purposes: to understand which assumptions and/or parameters influence model's results and to identify the region of parameter values where the central result of the model holds [86].

2.4 Limitations of ABM

With all its benefits over competing methods, ABM is not a panacea, though. As any other modelling method, it has its drawbacks.

The main drawback of ABM, which is also typical for other simulation methods, is the so-called *curse of dimensionality* [14]. Recall that ideally, we have to run the model for all values of parameters in the parameter space. The time required for this, however, increases exponentially with the number of parameters. In particular, as already noted, it is relevant for parameters that have to be calibrated. The model, thus, has to be kept compact enough to ensure that it can be calibrated without exceptional efforts and analysed in proper time (compare and contrast it with the requirement of analytical tractability of mathematical models). While ABM allows us to build models that are much more realistic than before, this issue reminds us that we are still operating in the field of constructing models as a *simplified picture of* reality and not in the field of *duplicating* reality.

Moreover, while agent-based models allow us to express the "enormous amount of data and knowledge about the behaviour, motivations and relationships of social agents" [22, p. 7199], there are concerns that we do not understand the mechanisms of human behaviour well enough to be sure that the models we build are correct [86]. Two comments should be added here. Firstly, in these circumstances, ABM can be used as a virtual laboratory to test different hypotheses of human behaviour we are unsure about. Secondly, ABM is extremely flexible, so when we increase our understanding of these mechanisms, we can easily incorporate them in agent-based models.

Finally, ABM is still a (comparably) young method of science. Consequently, there is currently no standard way to construct and describe agent-based models, as well as to analyse the data stemming from simulation runs. Furthermore, models are rarely comparable, as they entail highly heterogeneous theoretical content, explain different phenomena, and no formal tests are usually conducted to measure the relative performance of different models of the same phenomenon [106, 221].

Chapter 3
Using Agent-Based Modelling in Studying Labour–Education Market System

We're now ready for a discussion on how exactly one could build an agent-based model of labour–education market system (LEMS). This discussion will necessarily be quite abstract, because particular mechanisms built into the model (agents, their behaviour, interactions, other structures) depend heavily on the purpose of the model. I'll focus on general approaches and mechanisms that you may find useful when building agent-based models of LEMS.

Every model of LEMS should, by definition, contain two large elements—education market and labour market—and the links between them. Because networks are prevalent in reality, these should also be included in the model. After the details of the model have been decided on, perhaps the most difficult steps come—calibration and validation—and they typically come together (you want to set the unknown parameters to such values that your model is valid, right?). I'll pay particular attention to setting parameters based on micro- and macro-level data in the following discussion.

I will refer to two types of agents. The most important are agents representing individuals, which I will call *persons* for short. So when I write *persons* I mean agents representing individuals, and when I write *individuals* I mean real individuals (e.g., from empirical studies). The second most important are agents representing firms, and I will call them *firms* for short. This is an abuse of notation, but when unclear from the context I'll refer to either *firm-agents* (in the model) or *real firms* (in empirical data).

In this chapter, I will first discuss issues relating to modelling education market and then move to labour market. Modelling social networks is a large (and important) enough topic to be dedicated as a separate section. Finally, when the model of LEMS is built, some researchers argue that it should be obligatory incorporated in a larger model (to "close" the system). I close the chapter with discussing whether it should indeed be always done.

A. Tarvid, *Agent-Based Modelling of Social Networks in Labour–Education Market System*, SpringerBriefs in Complexity, DOI 10.1007/978-3-319-26539-1_3

3.1 Modelling Education Market

This is perhaps *the* element that distinguishes a model of LEMS from a model of labour market. There is a major difference between a labour-market model and a LEMS model in how they view the aspects of education. In the former case, persons have access to a very simple mechanism that raises their qualifications, which later are used in their jobs. In the latter, this mechanism is much more detailed to study the imperfections of the education market and its link to the labour market. The level of detail, of course, depends on the purpose of the study, but there are several common issues you should think about when creating the education-related part of any agent-based model of LEMS (see Fig. 3.1 for details).

These could be grouped in two big tasks: define the structure of the education system and define the behaviour of persons there. Agent-based models of higher education system[1] [135, 217] typically include the three-step process of entering an education institution. These steps are application, admission and enrolment. Interestingly, we'll see a very similar three-step process of finding a job in Sect. 3.2. The three-step process, however, assumes that agents already decided that they want to attain a particular education level, so it skips the considerations of the

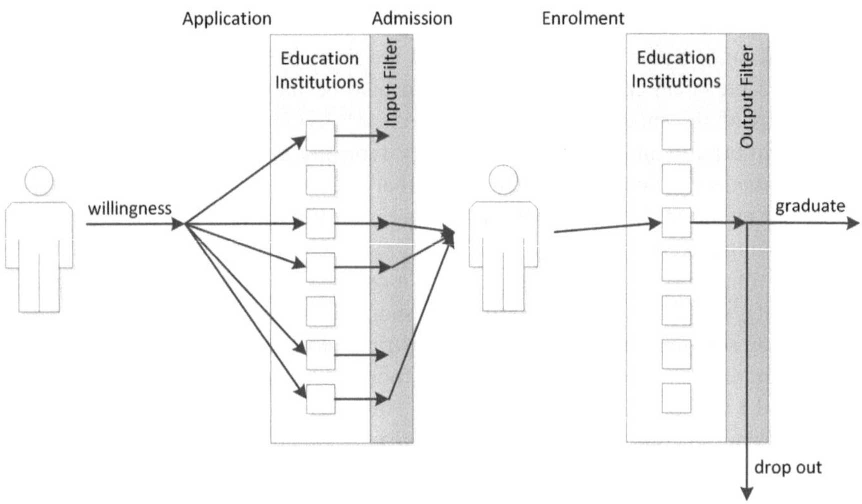

Fig. 3.1 Individual behaviour in the education market. The model includes the three-step process of entering a particular education level: application, admission and enrolment. In this illustration, the person applies to five education institutions, but only three reply that they admit it. The person then chooses one of these and either graduates or drops out

[1] See [126] for a more elaborate review of agent-based models of the education market.

person on the necessity of that education level, or its willingness to study, for short. Neither it includes the ultimate outcomes of the willingness to study: dropping out or graduating. All these steps are discussed in turn in this section.

3.1.1 Modelling the Structure of Education System

The structure of the education system in an agent-based model depends on both the structure of the actual education system that is modelled and the aim of the study. You may view the education system as a three-dimensional structure and then decide which dimensions to include and to what extent.

The first dimension is the level of education (typically called *vertical differentiation* of education institutions). It might be primary, secondary and tertiary. If primary education level is compulsory in a given country and its population generally attains it (which is typically the case in developed countries but not necessarily in developing countries) then, for simplicity, it makes sense to assume that all persons have primary education when they enter the model. Primary education might still be included in the model if it is necessary to model the time spent in primary education (e.g., if the whole life cycle of an individual is modelled), there is a substantial part of population that don't study in a primary education institution (or when a particular minority group weakly involved in primary education is modelled) or other dimensions are included at this level. The situation is the same with secondary level. Tertiary level typically should be included, because it's an important component of LEMS nowadays (recall, for instance, the problem of overeducation, which is largely connected exactly with tertiary education), and it could even be split into sub-levels, such as bachelors, masters and doctors.

The second dimension is the type of education (typically called *horizontal differentiation* of education institutions). At secondary level, it might be general versus vocational education, and these could be further divided by specialisation. At tertiary level, it might again be general versus vocational programmes, but there is an additional important parameter: field of study.

The third dimension is the number of agents representing each combination of level and type of education. There can be one agent or multiple agents. A single education institution represents the possibility of getting the corresponding level and type of education. If access to this education institution is primitive (e.g., with the same probability for all agents), the presence or absence of this education can be simply a property of the person. Otherwise, if access to it and its operations are modelled in more detail than that, or it is necessary to gather specific statistics about it, it makes sense to create a separate class[2] for this agent. If the heterogeneity of education institutions of a given level and type is important for the model, which

[2]Assuming that the agent-based model is implemented in some object-oriented language.

includes modelling the differences among them that go beyond the level and type of education, then multiple agents representing these education institutions should be included in the model.

Choosing a particular structure of the education system along these dimensions is not the end of the story. In a typical education system, there are two particularities that should be taken into account in the model. Firstly, there are well-determined pathways between the levels of education depending on the type of education acquired on previous level(s). For instance, graduates of a vocational secondary school are typically not allowed to enter an academic higher education institution without additional time spent in acquiring the knowledge they lack. Secondly, one education institution might cover several levels and types of education. Typically, higher education institutions provide not only bachelor-level, but also master's-level programmes, though not all provide doctoral-level programmes. Similarly, they typically provide programmes in several fields of study, although specialised institutions also exist. Both particularities could make the proposed three-dimensional view of the education system somewhat simplistic, but it is, nevertheless, useful. What concerns the second point, you could always create meaningful types of education institutions so that they could fit neatly into the view.

An illustration of this approach is given in Fig. 3.2. Persons could enter primary education, then go to either general or vocational secondary education institution, as shown by the arrows in the figure. Note that primary education and each of the types of secondary education are represented by a single agent, which in this example means that the details of the institutions of these level–type combinations are not important for the model. After graduating general secondary school, agents could

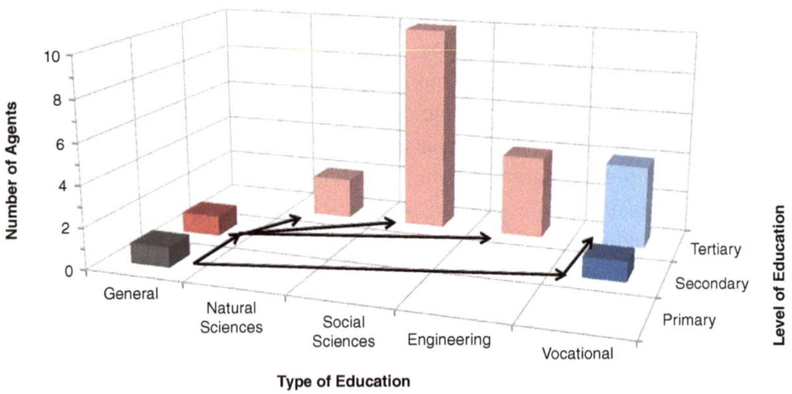

Fig. 3.2 Example of a 3D view of education system in an agent-based model. There are three education levels: primary, secondary and tertiary. At primary level, there is a single agent of a single type. After graduating primary education, persons move to secondary education, where they choose between two agents, representing general and vocational schools, respectively. At tertiary level, there are four types of education institution agents: three fields of study available to graduates of general secondary school and one type representing vocational higher education. There are several agents of each type at tertiary level

continue to three types (here, groups of fields of study) of higher education institutions: two institutions providing education in natural sciences, ten in social sciences and four in engineering. Graduates of vocational secondary school may choose to continue studies in one of the four vocational higher education institutions. Differences in vocational programmes are also not important in this model, as there is only type of vocational higher education institutions.

The settings in this illustration do not preclude you to actually make one (traditional) university providing education of all three types and another (technical) university providing education in both natural sciences and engineering. In this case, there remain eight institutions with programmes in social sciences and two providing engineering education. Consider changing the types of education at this level, so that they are better grasped visually.

Some would still argue that this three-dimensional view is incomplete, especially for models where a strong spatial component is needed. In other words, these models define where, in some geographical space, education institutions are located, so geographical space should be the "fourth dimension" of the education system structure. I don't support this view: even if the spatial component is important, it should be the institutional agent's property and not a "fourth dimension" of the education system structure. One of the reasons is that the 3D view I propose here is simple and easy to grasp but powerful in that it allows to overlay education paths on it. There are too few cases when education path depends on the spatial characteristic of previous education to overcomplicate the 3D view. That's why I would define the structure of the education system and its spatial distribution separately.

3.1.2 Modelling the Willingness to Study and the Application Step

Here, I will consider together two decisions: whether to study (motivation) and where/what to study (application), because they can be considered together as a problem of discrete choice among several options.[3] Technically, the person has to choose an option from set $F = \{f_0, f_1, \ldots, f_M\}$, where f_0 is the no-change state (i.e., "don't study"), while the other M options are fields of study or institutions or their combination, which I'll call "education paths".

Traditional economists would tell us here to define a utility function and postulate that the agent will choose the option with the maximum utility. This proposal is not without its merits, provided that we define the utility function in a way that keeps the

[3]The two decisions may depend on distinct sets of factors, but this is not a problem from the mathematical point of view.

heterogeneity of agent behaviour. One of the ways to do it is to make the function dependent on agent characteristics, including some characteristics of their social networks.

Then our agents could act in the following way:

1. Take each alternative f_i
2. Measure its utility $u(f_i)$
3. From all alternatives, choose the one with the highest utility:

$$f_i = \arg\max_{f_i \in F} u(f_i)\,. \tag{3.1}$$

This is a typical behavioural algorithm for a *perfectly rational* agent. If we want our model to be closer to reality, we should consider how to weaken the perfect rationality assumption and instead model *bounded rational* agents. Such agents *satisfice* [242] instead of optimising, searching "the available options just long enough to find one that reaches a pre-set threshold of acceptability" [68].

To model satisficing behaviour, we can tweak the above three steps in several non-exclusive ways:

- In step 1, consider a *subset* of alternatives instead of all alternatives:

$$f_i = \arg\max_{f_i \in \widetilde{F}} u(f_i),\ \widetilde{F} \subset F \tag{3.2}$$

  For example, choose only from the universities located in the city where the person lives.
- In step 2, measure the utility of alternatives imprecisely, with some error:

$$f_i = \arg\max_{f_i \in F} \tilde{u}(f_i),\ \tilde{u} = u + \varepsilon \tag{3.3}$$

  For example, add a normally distributed random error to the average of friends' opinions on the quality of university.
- In step 3, choose randomly from the alternatives with utility higher than a pre-specified threshold:

$$f_i = \text{random}\,\{f_i : u(f_i) \geq \overline{u}\} \tag{3.4}$$

  For example, choose only from fields of study where the expected wage after graduation is at least twice higher than average wage in the country.
- In step 3, choose randomly from the alternatives with utility sufficiently close to the optimal alternative:

$$f_i = \text{random}\left\{f_i : \max_{f_j \in F} u(f_j) - u(f_i) < \varepsilon\right\} \tag{3.5}$$

For example, choose only from fields of study where the expected monthly wage after graduation is not more than \$100 lower from the maximum expected monthly wage across all fields.

After we understand how persons determine whether and where to study, we should define the utility function $u(\cdot)$ using our knowledge of actual human behaviour described in Chap. 1. Recall that individual behaviour in the education market is determined by three groups of factors: individual peculiarities (e.g., inherent interest to engineering), social environment (family, social networks, social class, race, ethnicity etc.) and monetary costs and benefits (tuition fees, transportation, accommodation, financial support, expected income in the labour market, etc.). Two factors are added to these in [217]: whether they are desired by the education institution and whether, given that level of desirability, they will be accepted—in short, they also consider the probability of being admitted (which is logical if submitting an application has some costs attached to it).

Some of these factors are regressive—like the ability level or race and ethnicity—while some are clearly progressive: persons may aspire elevating their social class by getting a higher education level than is usual in their family or monetary benefits will not be felt until after (a few years after) graduation. Not all costs are immediate—there is a cost of foregone time, during which competitors in the labour market gain experience, while students gain knowledge—and these costs might not be realised by persons when deciding on the education path.

At this point, modelling becomes complicated in two ways. Firstly, all these groups of factors interact, so that the person might have higher abilities in a given, e.g., field of study, but due to the influence from social environment, it may choose a different field or not to study at all. It sounds like weighting should be applied, but it's completely unclear how to parametrise the weights or their distribution, unless the specific data on this issue are available. It then makes sense to leave the selection of parameters until the calibration phase of model building, which also means that the number of other parameters subject to calibration should be small.

Secondly, all these factors have to be computed. The difficulty here lies in the approach to assessing progressive costs and benefits. These should not necessarily be readily available to persons through some official statistical data portals or job placement centres of education institutions, but even if these are, they should not necessarily know about it or trust these data. For instance, the information on expected salary rise after graduating a programme available from the job placement centre of the education institution hosting the programme may be biased, because the education institution has an incentive to inflate it.

Take the estimation of an increase in income after getting a bachelor degree in a certain field. Assume that there are official data on that from one of vacancy publishing portals. There is also information that the unemployment rate among the graduates of that field is higher than the average among all tertiary graduates. Persons will not necessarily take that information into account. Besides the already mentioned lack of trust to the source of information, they may be overconfident. The former may manifest as looking for concrete examples of the actual income earned by graduates instead of drawing conclusions from some anonymous averages. For instance, persons will gather this information from their social networks. The latter (overconfidence) will, in this case, manifest as taking into account only the best indicators—e.g., average income over the upper half of the income distribution in the social network. Clearly, there are many variants how exactly this behaviour may be specified and how these specifications influence the whole LEMS, as amplified by the social network lens.

This example illustrates an important point that was raised back in Chap. 1 about the different criteria used in choosing the education path as compared to the criteria used to assess the fit of the person to the respective job. Call the latter criteria the $\mathcal{F}$-characteristics of the person. For instance, these may include mathematical skills, which are of low importance to a linguist, of medium importance to a programmer and of high importance to a physicist. Then a person with low mathematical skills should consider *not* studying physics, because the level of correspondence with the job will be too low. Or these may include some personality factors, e.g., one of the six Holland's personality types discussed in Chap. 1.

In principle, the utility function should be based on the level of correspondence of $\mathcal{F}$-characteristics of the person and the job (and several other factors, such as tuition fees). Actually, the decision is taken based on a set of $\mathcal{H}$-characteristics. These may, for instance, include friends' opinions (amplified by the general importance of friends' opinions to the person in this case). This is one of the ways how poor match between the person and the job might result: the person chooses the education path based on $\mathcal{H}$-characteristics, ignoring $\mathcal{F}$-characteristics, which actually determine success in the job (and, in particular, the level of income). In [260], the sets of $\mathcal{F}$- and $\mathcal{H}$-characteristics did not intersect, but this should not necessarily be the case.

Consider a few examples from existing agent-based models incorporating the behaviour in the education market. The willingness to enter university in [262] depends on the share of tertiary-educated friends of the person and on the type of the person, so that those with more friends who decided to enter the higher education market are in general more likely to also desire getting higher education. In the models concerning the secondary education market, persons choose the school with maximum product of average pupil achievement and geographical distance [180] or taking into account geographical distance, school performance (for specific social classes) and sex and/or religion (for specific schools) [129]. In higher education market models, persons choose institutions with average quality of students (as measured by SAT or GPA scores) closest to their quality [135] or the largest product of the utility of attendance (depends on the institution's quality) and the probability of admission (depends on the closeness of institution's quality and person's quality) [217]. In [217], agents measure their own quality and that of institutions with an error, using a variant of (3.3). Field of study choice in [260] is governed by five $\mathcal{H}$-characteristics of the person: share of friends who graduated the field, the average wage (calculated over the top half of the distribution) and the employment rate of friends who graduated the field, and the average wage (calculated over the top half of the distribution) and the employment rate of friends who work in the profession corresponding to the field. Note how the overconfidence of agents in comparing their employment prospects with the actual state of their friends was modelled.

Another question is the possibility of reconsidering the desire to study. Many individuals choose to stay in the labour market for some time after graduating secondary school instead of going into higher education institutions right from the school. Or, if drop-outs are modelled, whether persons are allowed re-entry into the education market after dropping out from it. Consider whether skipping these possibilities affects the results of the model.

3.1.3 Modelling the Outcomes of the Willingness to Study

Whether persons wish to enter studies in a certain institution and/or field and apply there does not guarantee that they will graduate. They firstly have to be admitted to the study programme, then they have to choose where to enrol, successfully pass all courses (and not drop out) and only then they will graduate. These correspond to input and output filters in Fig. 3.1.

The possible regimes of the input filter can be described using a two-dimensional view shown in Fig. 3.3. The education institution typically decides (within the limits allowed by the respective legislation, to which the model has to be calibrated) on the quantity and quality of each admission wave. The legislation was mentioned primarily because governments frequently require secondary schools to admit students

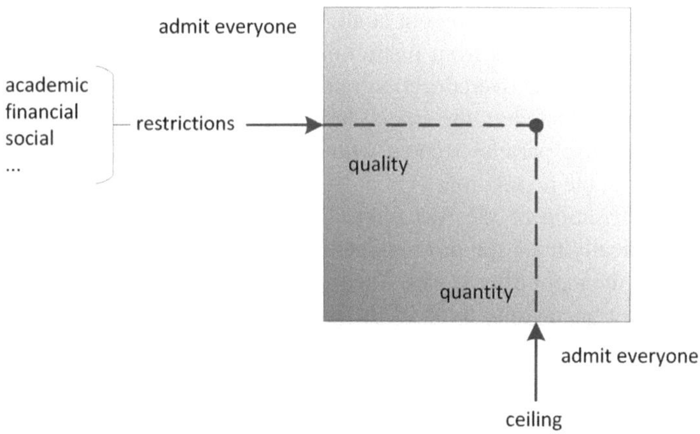

Fig. 3.3 A 2D view of input filter regimes. The illustration shows the decomposition of a given input filter regime (shown by a *point inside the square*) into quality restrictions and ceiling on quantity. The *lower left corner* symbolises maximum restrictions, and the *upper right corner* is the "let everyone in" regime

living in their area if they wish to study there, regardless their academic ability. In modelling European higher education system (ignoring elite universities), a frequent regime of the input filter is to pass everyone [260, 262] up to a certain ceiling [262]. "Everyone" here still means that the applicants graduated secondary school with minimum qualifications required by state examinations (or their equivalent), of course. Nevertheless, institutional agents in the models of higher education may decide to admit only the highest quality students up to a ceiling [217] or first sort the applicants on their ability (SAT/GPA scores in the US context) and then take half of target admission size above the (exogenously determined) threshold and half from below it [135]. Less elaborate filter structures were modelled in secondary education: based on the first-come-first-admitted (up to a ceiling) basis [180], to which an additional filter of sex and/or religion could be added to model specific types of schools [129].

The output filter, in contrast, does not have the quantity side and can only force students out of the institution based on their quality.[4]

The enrolment process is modelled, for obvious reasons, only if persons can apply to several education institutions and can be based on some measure of the quality of the institution (the utility of attendance in [217] or a function of the similarity between the person and the institution's students, costs and geographical proximity of the institution in [135]).

The behaviour in the education market is complicated by the parallel existence of the labour market (as was just mentioned in case of reconsidering the desire to study). Students may easily be able to work in parallel, and you have to consider whether allowing this possibility in the model changes its results. It should not affect the results if these are primarily related to the distribution of the education level in the population and (1) parallel employment does not affect the probability of dropping out (i.e., the probability of graduating at all or graduating on time) or (2) there is no possibility to drop out at all. This is the case, for instance, when the model simulates the effects from restricting the accessibility of higher education system depending on the overeducation rate without a possibility to drop out from university [262]. In that case, the crucial factor was actually the lack of mechanism to drop out, because parallel studies without it do not affect the distribution of education level in the population.

The parametrisation of the two filters should not be a problem, as education institutions typically track the drop-out statistics, and it is easy to obtain the information on how the input filter works. The parameters should also not necessarily be constant—e.g., a linear time trend of the participation rate and the dependence of graduation probability on age are assumed in [168].

[4]Strictly speaking, in case of private education institutions, the quantity side still plays a role here, as they may be reluctant to turn the output filter on to full extent if that would mean such large drop-out rate that would make them miss their financial targets.

3.2 Modelling Labour Market

When modelling the labour-market aspects of LEMS, researchers are typically interested in the behaviour of persons. The presence of firms is typically coupled with some sort of vacancy mechanism, and it immediately adds another complicated layer to the model. I will consider both cases here, starting with a seemingly simpler case of person-only model.

3.2.1 *Skipping Firms and the Vacancy Mechanism*

Why skip firms and the vacancy mechanism if we're already writing an agent-based model, which can be as complex as we wish? After all, we aren't constrained by mathematics, so let's write a full-scale LEMS model! This is typically the first thought of beginners. The answer to this question is not that straightforward.

Remember that for any research goal, the model should be as simple as possible but, nevertheless, allow to perform a thorough study. Adding a layer of complexity is useful if it pays off. We'll see many examples of that later on. In general, a standard economic argument applies here: it makes sense to add complexity to the model until the marginal costs of complexity become higher than its marginal benefits.

In a model where there are only persons in the labour market, researchers have to build a probabilistic model of the latter that allows to simulate the observed dynamics. Consider the most general scenario. There are n types of agents and m types of jobs.[5] The probabilistic structure would then be a collection of probabilities of finding job p_{ij} and losing job (or quitting) q_{ij}, where $i \in \{1, \ldots, n\}$ denotes agent type and $j \in \{1, \ldots, m\}$ is the current (in case of losing job) or the last (in case of finding job) job.

How can we parametrise this probabilistic structure? Assume we have micro-level data where we can identify all n agent types and m job types (if we can't, we'll have to study the model across several reasonable values; in most cases, we either can do it or we change the model or the dataset). Let's start with a simple case of no temporal dependency. I'll suppress indices i and j for readability.

Proposition 3.1 (Setting Time-Independent Probabilities to Find and Lose Job). *Let u be empirically observed unemployment rate and k average length of unemployment spell (in time units equivalent to model periods; e.g., 4 months if one model period represents 1 month). Then, if the probabilities to find and lose job in the model, p and q, respectively, are time independent, they should be set to*

[5]There's no mistake here—it is possible to introduce job heterogeneity without introducing firms. This might, for instance, reflect occupations or industries.

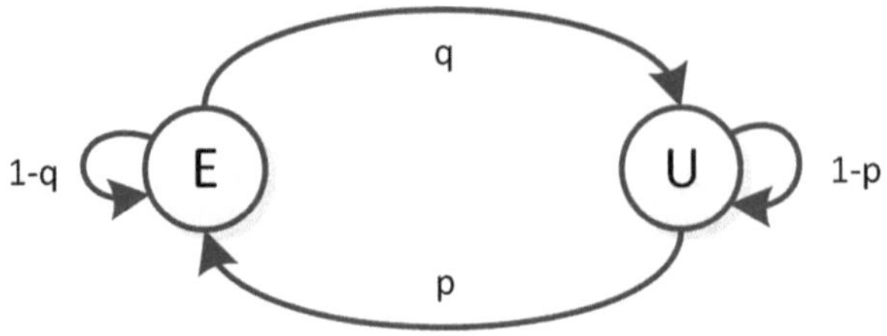

Fig. 3.4 Two-state Markov chain of balancing between employment and unemployment. E and U are employment and unemployment states, respectively, and p and q are the probabilities of finding and losing job, respectively

$$\begin{cases} p = 1/k \\ q = pu/(1-u) \, . \end{cases} \tag{3.6}$$

Proof. Without time-dependence, we're operating with a memoryless process. In other words, the set of probabilities determining the next state in which the person will be depends only on its current state; previous states are irrelevant for it. This process can be modelled as a two-state Markov chain, as shown in Fig. 3.4. States E and U represent employment and unemployment, respectively, and probabilities p and q guide the transitions between them.

In the steady-state, the probabilities of being in states E and U, π_E and π_U, respectively, are given by the following relationships:

$$\begin{cases} \pi_E = (1-q)\pi_E + p\pi_U \\ \pi_U = (1-p)\pi_U + q\pi_E \\ \pi_E + \pi_U = 1 \, , \end{cases} \tag{3.7}$$

from where we get the relationship between transition probabilities:

$$q/p = \pi_U/(1-\pi_U) \, . \tag{3.8}$$

It remains to set $\pi_U = u$ to get the second line of (3.6).

Equation (3.8) allows us to constrain only the ratio of p and q, but not their actual values. Assume that $u = 4\,\%$, so that $u/(1-u) = 1/24$. What is then the difference between model A with $p = 6\,\%$ and $q = 0.25\,\%$ and model B with $p = 72\,\%$ and $q = 3\,\%$? Both will have unemployment around $4\,\%$, but model B will clearly be much more volatile in the sense that persons there will quit jobs around six times more frequently than in model A (they will also be able to find jobs around six times faster, so that the unemployment rate doesn't change).

It is then clear that the values of p and q should be set according to the length of staying in employment or unemployment states. Statistics on average unemployment spell length are typically easier available than that on average employment length. Provided that when the person enters the unemployment state, it stays there for at least one period, average length of unemployment spell can be written as

$$1 + (1-p) + (1-p)^2 + (1-p)^3 + \cdots = \frac{1}{1-(1-p)} = \frac{1}{p}. \tag{3.9}$$

The first line of (3.6) follows from equating (3.9) with its empirically observed equivalent, k. □

There are two takeaways from this proposition. Firstly, even in the seemingly trivial case of time independence and two probabilities, knowing only the target unemployment rate is insufficient to set parameter values. Secondly, calibration could be performed to find the necessary value of p from which to deduce q, but you can save much time if you can do the necessary (pretty simple in this case) calculations by hand and use your time instead to calibrate other parameters, those that are neither readily available in empirical data nor can be deduced from there.

Consider now a more difficult, and more realistic, case of parametrising a probabilistic structure with temporal dependency. For instance, as it was discussed in Sect. 1.3, the probability to find a job (or, equivalently, exit unemployment) depends on the time already spent in unemployment, and this might be reflected in the model by imposing a third dimension on parameter p, making it p^t_{ij}, where the superscript t denotes time spent in (the current) unemployment spell.

The relevant empirical data are typically available in the form of tabulation of the share of the sample with unemployment spell of 1 month, 2 months, 3 months etc., which have to be converted to the respective probabilities. The probability to lose job, q, as you might guess from Proposition 3.1, should be related to p to generate the empirically observed unemployment rate, but the relationship is not as straightforward as in (3.6). The following two propositions analyse this case.

Proposition 3.2 (Setting Time-Dependent Probability to Find Job). *A temporal structure in the data is given by a set of shares of respondents with different unemployment spells,* $\{f_t \,|\, 0 \le f_t \le 1 \wedge \sum_t f_t = 1\}, t = \overline{1,T}$. *Given these, the probability to exit unemployment after spending t periods there is*

$$p_t = \frac{f_t}{1 - \sum_{k=1}^{t-1} f_k}. \tag{3.10}$$

If the last share, f_T*, refers to an open interval (i.e., the share of respondents who found a job after T or more periods) and it is assumed that* $p_T = p_{T+1} = p_{T+2} = \ldots$ *then* $p_t \equiv 1/2$ *for all* $t \ge T$.

Proof. Assume that in period 0, the person lost job. In period 1, the person can find a job (exit unemployment) with probability p_1 or not find it (continue searching in period 2) with probability $(1 - p_1)$, and similarly in periods 2, 3 and so on. Empirically, we observe f_1, the share of persons with unemployment of length 1 period, f_2, the share of persons who found job after 2 periods, etc. The key to relating $\{p_i\}$ with $\{f_i\}$ is to understand that f_i is actually the probability that the person did not find a job *before* period i but found it *in* period i:

$$f_i = (1 - p_1)(1 - p_2) \dots (1 - p_{i-1})p_i \,. \tag{3.11}$$

The expressions for p_i are then obtained inductively:

$$p_1 = f_1 \tag{3.12}$$

$$p_2 = \frac{f_2}{1 - p_1} = \frac{f_2}{1 - f_1} \tag{3.13}$$

$$p_3 = \frac{f_3}{(1 - p_1)(1 - p_2)} = \frac{f_3}{1 - f_1 - f_2} \tag{3.14}$$

$$\dots .$$

Now assume that the last empirical observation, f_T, is the share of those exiting unemployment after T or more periods and $p_T = p_{T+1} = p_{T+2} = \dots = p$. Denote the probability to stay unemployed for $(T - 1)$ consecutive periods by P. Then, assuming infinite life of a person, we have the following relationship:

$$\begin{aligned} f_T &= Pp + P(1-p)p + P(1-p)^2 p + \cdots \\ &= P \sum_{i=0}^{\infty} (1-p)^i p \\ &= P \frac{p}{1-p}, \end{aligned} \tag{3.15}$$

from where we get

$$p = \frac{f_T}{f_T + P} \,. \tag{3.16}$$

It is easy to see that this expression is equal to $1/2$. Indeed, consider an event that the agent finds a job in period T, $T + 1$ or later. The probability that this event happens is f_T. But the fact that this event happens is equivalent to the event that in the first $(T - 1)$ periods, the agent was unemployed. The probability of this event is P. Hence, $P \equiv f_T$ and (3.16) evaluates to $1/2$. □

Proposition 3.3 (Setting Time-Independent Probability to Lose Job Given Time-Dependent Probabilities to Find Job). *Given a set of time-dependent probabilities to find job, $\{p_i\}$, and the unemployment rate, u, the time-independent probability to lose job is*

$$q = \left[1 + \sum_{i=1}^{T-2} \prod_{j=1}^{i} (1 - p_j) + \frac{(1 - p_1) \dots (1 - p_{T-1})}{p_T} \right]^{-1} \times \frac{u}{1 - u}, \tag{3.17}$$

where T is the spell length such that $p_T = p_{T+1} = p_{T+2} = \dots$.

Proof. Consider again the two-state Markov chain from Fig. 3.4. Recall from Proposition 3.1 that the following steady-state relationship between transition probabilities holds:

$$q = p\pi_U/(1-\pi_U)\,, \tag{3.18}$$

where π_U should be set to the empirically observed unemployment rate, u.

The problem is that we do not have a single probability of exiting unemployment, p, but instead we have a set of probabilities depending on the length of staying in unemployment, $\{p_i\}$. One of the possible approaches to tackling this problem is to set p to the value that would give an equivalent average unemployment spell length as $\{p_i\}$, i.e.,

$$1+(1-p)+(1-p)^2+\cdots = 1+(1-p_1)+(1-p_1)(1-p_2)+\cdots\,. \tag{3.19}$$

Both expressions will give the average unemployment spell length, which was denoted as k in Proposition 3.1.

For compactness of the following calculations, denote

$$\lambda_i \equiv (1-p_1)(1-p_2)\ldots(1-p_i)\,. \tag{3.20}$$

Because in empirical data, we typically have spell length T such that $p_T = p_{T+1} = p_{T+2} = \ldots$, we can break the right-hand side of (3.19) into two sub-expressions:

$$1+\lambda_1+\lambda_2+\cdots+\lambda_{T-2} = 1+\sum_{i=1}^{T-2}\prod_{j=1}^{i}(1-p_j) \tag{3.21}$$

and

$$\lambda_{T-1}+\lambda_{T-1}(1-p_T)+\lambda_{T-1}(1-p_T)^2+\cdots = \frac{\lambda_{T-1}}{1-(1-p_T)} = \frac{\lambda_{T-1}}{p_T}\,. \tag{3.22}$$

Plugging both parts back into (3.19) and rewriting the left-hand side of the equality, we get

$$\frac{1}{p} = 1+\sum_{i=1}^{T-2}\prod_{j=1}^{i}(1-p_j) + \frac{(1-p_1)\ldots(1-p_{T-1})}{p_T}\,. \tag{3.23}$$

It remains to substitute the resulting expression for p in (3.18). □

3.2.2 Adding Firms and the Vacancy Mechanism

In many cases, a simple probabilistic structure of bouncing between the employment and unemployment states is not sufficient to reach the goal of the study. You might consider adding vacancies or even firms to model labour market dynamics more closely.

Hold on, you'd say, why are you writing "vacancies *or even* firms?" Shouldn't firms be *always* included as soon as vacancies appear in the model? The answer is no. Look at it this way. What is the difference between a vacancy and a firm? A vacancy is simply a specification of (1) the job where some person will work in a few model periods (unless the vacancy is deleted as unfilled) and (2) the (minimum) qualifications of a successful candidate. So if all you need in the model is a process generating vacancies with certain attributes, then you'll just have vacancies in your model and no firms—because if you do include firms, you'll have to add a bunch of parameters for firms, waste time on setting these parameters (in the worst case, calibrating them) and, perhaps even worse, waste your readers' time on the description of the set-up of the firm component in the article where you publish your work. Remember that the model should not be unnecessarily complex.

In modelling LEMS, it is useful to consider firms as collections of jobs and, more importantly, persons. Looking from this angle, it makes sense to include firms in three cases. Firstly, if you have to model entrepreneurial processes—how firms are created and how they die. Secondly, if you have to model human resource practices—how vacancies are created, how employees are attracted and hired, how employment relations change (e.g., promotion mechanisms) and how employees are laid off or when they quit (including bulk lay-offs when firms die). Thirdly, if employees' decisions depend on co-workers (i.e., on on-the-job social networks) or on job embeddedness in general. In any of these cases, firms should be a well-defined construct in the model.

Recall the three-step procedure of application, admission and enrolment in the education market shown in Fig. 3.1. We can model individual behaviour in the labour market along the same three steps: application for a vacancy, job offer (by the firm to the successful applicant) and choice among offers (in case of receiving several offers). Similarly, just as the education process ends with graduation, work relationships end with separation, either voluntary (job quit) or involuntary (lay-off).

It's instructive to look how existing agent-based models implemented the vacancy mechanism. They all can be represented as points in the 3D space, depending on person behaviour (how many applications are sent and how the offer is selected if several applications are successful) and firm behaviour (how the successful applicant is chosen), see Fig. 3.5 for an illustration.

The simplest case to model both person-side choices is to allow firms to pick from the unemployed who passively wait [1, 257]. This is equivalent to a person applying to all vacancies but receiving the decision to be accepted from no more than one firm, so that the person does not have to choose from firms after applying.

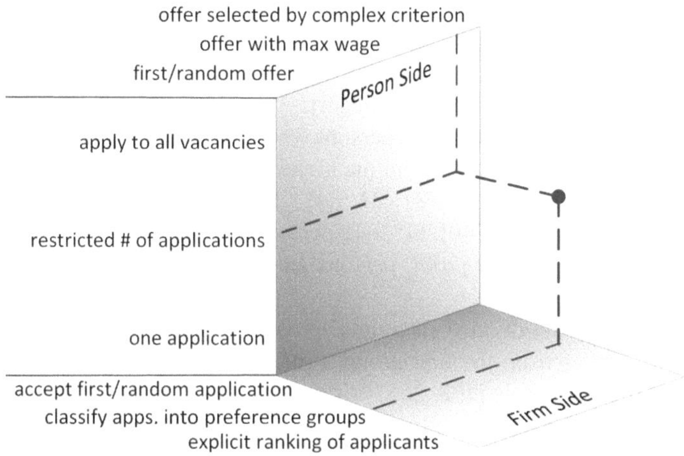

Fig. 3.5 A 3D view of the vacancy mechanism in agent-based models. The *point* represents a model in this 3D space. It projects on the plane representing person behaviour, where it determines the number of applications sent (here, some restricted number of applications) and the method of selecting the offer (here, persons use a complex criterion to rank offers). It also projects on the plane representing firm behaviour, and there it determines how an application is chosen (here, by classifying applications into preference groups and then picking a random application from the most preferred group)

There are models where persons apply for all vacancies [117] and the variations of it: all vacancies in their neighbourhood [181], all vacancies firms probabilistically inform them about [266], all vacancies they find out through social networks [173, 266, 270] if the proposed wage exceeds their reservation wage [80] or all vacancies in the industries matching their investments in human capital [200]. A variant of that is also applying to a random set of firms [80], because the number of applications is not bound from above.

On the other side of the spectrum are models where persons select only one vacancy to apply for or one firm where to search for vacancies. Persons may apply for a single vacancy with the maximum utility exceeding the reservation utility [20, 21], choose to search for vacancies in a single firm chosen probabilistically [105] or firstly contact the last employer and choose a random firm only if the last employer does not have vacancies corresponding to the type of the person [240]. One could also include in this group studies where persons do not search for vacancies, but are informed about them by firms always [202] or probabilistically [265] or by their friends [265]. In both [202] and [265], persons apply for the vacancy with the highest wage, but in [265], employed persons apply only if it is higher than the wage they currently earn.

Somewhere in the middle are models that put a cap on the number of vacancies a person could apply for. Typically, these are randomly selected vacancies [220] with wages exceeding the reservation wage [81, 82, 84, 87]. Alternatively, these can be the vacancies with the highest utility—e.g., the highest expected job satisfaction [264]. Although not modelling vacancies as such, persons in [268] send work offers to a number of firms (limited from above) with the highest non-negative utility, which puts this model in this middle category.

In case of receiving several offers, persons typically decide based on the proposed wage only: they choose the offer with the highest wage [80, 87], sometimes taking into account commuting costs [81, 82, 84], if the proposed wage exceeds the reservation wage [105, 220, 240]. In [264], persons choose the offer with the highest expected job satisfaction. In [200], persons simply accept the first job offer they receive.

Studies can also be divided into three groups based on how the firm-side is modelled. In studies from the first group, firms choose the applicant on the first-applied-first-accepted basis [202, 265] or through purely random choice [181, 200, 266],[6] in some models not accepting the applicants it has just laid off [173, 270]. Some studies do not explicitly state whether the first or a random applicant is chosen [105, 240], but it is otherwise clear that one of these mechanisms should be used: the firms there just propose a wage for a vacancy and accept whoever agrees with that wage offer. Always specify all details of your agent-based model; it's considered *mauvais ton* if you don't.[7]

In studies from the second group, firms classify applicants into few distinct groups, preferring some groups to others, and then choose persons inside the non-empty group they prefer the most. Models using referral hiring typically fit in this category. In [1], firms probabilistically choose whether to use formal channels or hire through referral. In the former case, they choose a random applicant, while when hiring through referrals, the firm has a higher probability of choosing the person of the same type as it has itself. In [264], in line with empirical evidence, only small and medium-sized firms may choose to use referral hiring, preferring friends of their current employees, while large firms always use formal search channels. Similarly, in [117], firms first choose randomly from the applicants having friends among its employees, and in [80], instead of random choice persons with

---

[6]These two variants are equivalent; the difference is only in the side where random choice is implemented. In the first-applied-first-accepted approach, the programmer randomises the sequence of persons in the queue of sending applications. In the random choice approach, persons send applications in any order (e.g., in the order they are stored in the array of labour market participants), and randomness is shifted to the process of choosing an element from the list of applications received by the firm.

[7]Besides leaving the reader somewhat disappointed by hiding the details of your model, you make it harder to replicate it. There is a strand of research on *agent activation*, which studies in which cases seemingly minor differences in the implementation details of the mechanism of agent behaviour cause substantial differences in model outputs (see, e.g., [15, 69, 70] if you're interested). Thus, a replica of the model may produce results that are inconsistent with the original model also if the replica's author incorrectly inferred the actual—but unspecified—activation scheme used in the original model.

the highest general skills are chosen. Firms in [257] use a five-step procedure, moving to the next step if the previous failed to fill the vacancy: (1) re-hire the previous employee with the highest quality if its quality meets standards and exceeds group reputation scores; (2) randomly hire an employed friend *of the firm*; (3) hire the previous employee of a business partner with the highest quality if its quality meets standards and exceeds group reputation scores; (4) hire a friend of recently employed workers randomly or in the order of the quality of referents; and (5) randomly hire an unemployed from the group with higher reputation.

Studies from the third group model an explicit ranking mechanism firms use to sort the list of applicants and then choose the applicant with the highest rank. Studies then differ with respect to which ranking mechanism they use. In some, firms use general skill level [84, 85, 87], and if persons have the same general skill level they are sorted by specific skill level [81, 82]. In others, firms choose the applicant with the highest expected profit above the minimum tolerated profitability of the vacancy [20, 21]. The firms in [220] employ the most productive applicant. Finally, while vacancies are not explicitly modelled in [268], firms keep the waiting list of applicants with the highest non-negative utility, the length of the list being limited from above, and update the list after every wave of applicants; firms employ everyone from the list if it receives no more applications, thus, employing the applicants with the highest utility.

One important moment was left out of scope—what to do with the vacancies that were *not* filled? Consider what a real firm would do in this case. It would either liquidate the vacancy or keep it for some more time; in the latter case, it would keep it unchanged, increase the proposed wage, decrease requirements or combine the latter two changes. After certain time, if the vacancy was still unfilled, it would be liquidated. Agent-based models of labour market also simulate firms as acting in one of these ways, depending on their specifics. The requirements (e.g., for minimum experience, education or skill level, etc.)—even if modelled—are rarely changed, however. The wage proposal of a failed vacancy, in contrast, is frequently adjusted—e.g., in [85, 240, 264].

Besides these three dimensions, agent-based models of labour market may include two additional elements. Firstly, it is on-the-job search, where the behaviour of persons is different from the unemployed job seekers in that they typically consider fewer vacancies—only those that are better (in terms of higher wage [265] or higher expected job satisfaction [264]) than their current job. Secondly, it is the implementation of some of the reasons behind the dependence of the probability to find a job on unemployment spell length. Where included, it typically takes the form of the dependence of reservation wage on time with same elasticity φ [80–82, 84, 85, 87, 264],

$$w_{t+1}^r = \varphi w_t^r , \tag{3.24}$$

with a random decreasing factor [105, 240] or in a more complex fashion [20, 21].[8] In case the reservation wage drops with constant elasticity, it is easy to set the elasticity parameter φ based on existing empirical studies.

In most studies, the decision to quit is not modelled and the separation of working relationships occurs only through the firing mechanism. Other studies may be divided into those modelling it purely probabilistically and those defining a more elaborate decision function of the person. In the former group of studies, the probability was set globally [85] or depended on person's homophily level and workplace composition in terms of person types [1]. The model in [268] should also be put in this group: while it assumes that persons quit because their employer did not provide them the benefits they expected to receive, this behaviour of firms is governed by playing a prisoner's dilemma game. In the latter group, the persons that quit are high-quality workers who are not promoted for a long time or have the utility of quitting that exceeds the utility of continuing work [20, 21] or the employees whose job satisfaction fell below the critical level and who found a better job through on-the-job search [264].

3.2.3 *Modelling Firm Dynamics*

The education market is a comparably stable market, in the sense that there are comparably few education institutions, which are created and closed down relatively infrequently. In contrast, the labour market is a much more dynamic market, which contains many firms, which are born and go bankrupt much more frequently. Hence, while it perhaps does not make sense to model the birth and death of education institutions (unless the research question is closely related to that and the time span studied is substantial), it seems to be quite practical to include some mechanism guiding labour demand at firm-level and, hence, the mechanism of the creation and destruction of firms. These are the two issues I'd like to discuss now.

In the following discussion, I will necessarily keep the focus on the micro-level, as I'm most interested in modelling individual behaviour in this book. I will, thus, ignore some models that include firm generation and destruction without modelling individuals.

The mechanisms of firm creation and destruction in labour market models are typically quite simplistic. In most models, the number of firms is fixed, but it can be achieved in two ways: either firms operate in a way in which they cannot exit the model [117] or they can go bankrupt and exit the model but then the same number of firms is created immediately. In the latter case, a firm is destroyed if it has no

[8] In [105] and [240], the authors actually consider reservation wage to be constant and reflect the absolute minimum wage under which the person will never be ready to work. Instead, they impose a random decrease of the "satisficing" wage—the one the person would like to receive—with unemployment duration.

workers [21], it has negative profits in a single period [105] or accumulated over a concrete number of periods [240], it has negative net worth [85, 219, 280] or a shock hits the sector where the firm operates [200]. If more detailed firm characteristics are modelled, new firms in this case are created with these characteristics set at, or close to, current average level [105, 240]. In [85, 280], new firms are not created, but the failed firms simply resume operations after a certain inactivity period.

In case the number of firms is not fixed, the conditions of firm destruction are still the same: no workforce [220, 264], negative profit [220] or negative net worth [202]. Entrepreneurship may not necessarily be modelled even in these models. For instance, in [202], a *fixed* number of new firms appear in each period, and these firms attempt to copy the characteristics of the most successful firm at the moment. Otherwise, agents may become entrepreneurs probabilistically [264] or if the expected profit from a start-up is higher than the expected income from working in an existing firm [220]. The parameters guiding the creation of new firms have to be carefully calibrated, as they cannot be as easily inferred from empirical data as the probabilities of finding and losing jobs discussed earlier in this section. The mechanism of becoming an entrepreneur and starting a company is, thus, still a relatively weakly developed part of LEMS models.

In contrast, many processes guiding labour demand were proposed. Logically, the level of details depends on how detailed firm's operations are modelled. The most straightforward way is to set a concrete level of vacancies to be published by every firm [200], choose labour demand from a given interval [117] or choose the change in size randomly from a given distribution depending on previous size and change in size [264]. The chosen change should not necessarily be implemented to full extent in the same period. For instance, firms in [264] distribute the chosen change in size for the year evenly through the 12 months, while firms in [220] close a pre-defined share of the gap between the desired and actual sizes in each period.

In an arguably less realistic setting where a position has a fixed one-period duration, firms increase or decrease the number of new vacancies by a random factor depending on whether the previous period's profits increased or decreased, respectively [105, 240].[9]

Labour demand has also been modelled in more detail from the cost-based or production-based views. The cost-based view assumes that the firm will take that many persons whom it will be able to pay [105, 219]. The production-based view assumes that the firm will set its target level of production and then publish new vacancies or lay employees off, depending on whether there is a lack or a surplus of the labour force, respectively [85, 280]. A combination of both views is [21], where a vacancy for a position is opened if expected profit from the position is positive, the vacancies that might cause losses due to potentially weak demand are closed and

[9] This setting does allow to model overall increase and decrease in labour demand without the need to specify which employees firms lay off in case of downsizing. A downside is that job search takes place every period, which might be too frequent for your purposes.

employees are laid off in case of overall losses in the amount at which the firm can regain profitability.

The criteria might be different for start-ups. For instance, in both [220] and [264], a start-up chooses its desired size, but start-ups in [264] choose only from the sizes of small companies and publish all vacancies immediately, rather than smoothing vacancy publication over the whole year, as "older" firms do in that model.

Firm-level accounting databases may be used to calibrate—or validate—the specification of labour demand in the model. For instance, Amadeus[10] database is used in [264] to set probabilistic firm behaviour of increasing or decreasing its labour force. These data can also be used to validate the distribution of firm size, as done in [264]. Nevertheless, you have to be cautious when using such databases and ensure that firm-level statistics were measured in the economic climate that's relevant to your model. For instance, if you're modelling firm dynamics in normal economic conditions, you should consider filtering out the periods of economic crises or recessions in your empirical data.

Firm behaviour may be modelled at various depths and levels of detail. In any case, however, the resulting macro-level dynamics have to be validated. The use of stylised facts was noted in Chap. 2 when talking about indirect calibration. Besides helping to calibrate the model, they also help in validation (actually, because the indirect calibration approach claims that you should calibrate the model in such a way that it generates certain stylised facts, whether your model generates stylised facts or not reflects its validity or lack of it, respectively). The following stylised facts should generally be shown by a model with firm behaviour:

- *Beveridge curve*—a negative non-linear relationship between the vacancy rate (the number of unfilled vacancies as per cent of labour force) and unemployment rate
- *wage curve*—a negative relationship between unemployment rate and wages
- *Okun curve* (reflects Okun's law)—a negative linear relationship between changes in GDP and changes in unemployment rate (makes sense to include only if it's possible to measure an equivalent of GDP in the model, of course)
- *Phillips curve*—a negative relationship between changes in wages and unemployment rate

3.3 Modelling Social Network

The generation of social networks is a very particular topic. On the one hand, social networks affect individuals' and, thus, also firms' decisions in LEMS, as we saw in Chap. 1. Hence, it would be strange if this influence of networks on micro-level

[10]Amadeus is a product of Bureau Van Dijk company and contains internationally comparable financial data on around 21 million European companies. For a global version, consider Orbis, a product from the same provider with data on more than 160 million companies.

decisions didn't affect the macro-level behaviour of LEMS. This is a good argument for including social networks in LEMS models, as, otherwise, we might be missing some its important characteristics. On the other, as in many other cases, calibration presents a problem.

But first we need to understand how to generate social networks. Theoretically, there are two options: (1) don't bother about technical details and use existing software to generate a graph with specified characteristics and (2) write the necessary code yourself. The first option is appealing because it doesn't present any difficulties (after understanding which software to use[11] and how to work with it). It is safe to use it if the aim is to generate some network, which will not change during the modelling process or will change minimally. The precise algorithm that builds the network needed is also not that important in this case, in the sense that it can be used as a black box. The only thing we care about is that the structure of the network conforms to our assumptions about it.

If the task is to model an evolving network, then chances are that you'll have to implement the algorithm and write the code behind network generation yourself. In this case, you'll have to think about using the network generation algorithm that is realistic, and not only that generates the kind of network you need. Reviewing all models is again out of scope here—in part because it would take at least one such book to analyse them, but also because social network analysis is rapidly evolving and new algorithms are proposed quite frequently. If you're interested, see [201, Chap. 10] and [8, 39, 96] for reviews of some dynamic scale-free social network generating algorithms.

Consider an example of a poor choice of an algorithm. You're modelling how the person who was just admitted to university makes friends with other students. You want to use an algorithm that generates scale-free social networks. One of these algorithms, a variant of the preferential attachment model analysed in depth in [67, Chap. 3], works as follows. With probability p, a new vertex v is added to the graph and connected to a random vertex u, otherwise, a new edge $\{r, s\}$ is added, where r and s are existing vertices chosen randomly with probability proportional to their degrees. This model does generate a scale-free network with exponent $\beta = 1+2/(2-p)$, so that $\beta \in [2, 3]$ because $p \in [0, 1]$. However, if the model period is 1 year and a new student enters university, he can only make one new friend with probability p—otherwise, he will live without friends for the whole year, which is not the most realistic scenario.

Choosing an appropriate algorithm is not the end of the story. You'll have to incorporate it in the model. It's quite likely that in this process, the algorithm will change and generate a network with somewhat different structure than you expected.

[11] For instance, the YANG external network generator is used in [173, 270], but network generating packages in Java or other general-purpose programming languages can also be used—an attractive option if the model is also created in that language (or in a framework that uses that language).

I'll provide an example in the following subsection. Then I'll move to the questions of modelling tie creation and separation and the calibration of the network part of an agent-based model of LEMS.

3.3.1 *Adapting a Network Generation Algorithm*

So the network generation algorithm was chosen, and now it's time to incorporate it in the agent-based model of LEMS. This may not be as straightforward as it sounds. Let's take one of the models for generating scale-free networks and study it as an example.

One of the algorithms that can be used to generate a scale-free network is Duplication model $\mathcal{D}(\rho)$ analysed in [67, Chap. 4]. This model depends on one parameter, ρ, and proceeds in two-step iterations. At the first step, called *vertex duplication*, a new vertex is added to the graph and connected randomly with an existing vertex. At the second step, *edge duplication*, the algorithm goes over each neighbour of the existing vertex and connects it to the new vertex with probability ρ. Thus, ρ is the probability that the new agent creates a connection with a friend of its new friend. The pseudo-code of Duplication model is shown in Algorithm 3.1.

It was shown [67, p. 78] that the model generates scale-free networks with the exponent β that is a solution to the following equation:

$$1 + \rho = \rho\beta + \rho^{\beta-1}\,. \tag{3.25}$$

Thus, in case of using Duplication model, one should first determine the target value of β (or its target interval) and then use (3.25) to determine the value of ρ to be used in the model. Figure 3.6 shows a graph with 100 vertices generated from scratch by running Algorithm 3.1 with $\rho = 0.45$ 100 times. The size of 100 vertices was chosen purely for visual purposes.

Algorithm 3.1 An iteration of Duplication model $\mathcal{D}(\rho)$

Require: $|V| > 0, \rho \in (0, 1)$
$V \leftarrow V \cup \{v\}$
  {Vertex duplication step:}
$u \leftarrow$ random vertex from $V : u \neq v$ **and** $\{u, v\} \notin E$
$E \leftarrow E \cup \{\{u, v\}\}$
  {Edge duplication step:}
for all $w : \{w, u\} \in E$ **and** $w \neq v$ **and** $\{w, v\} \notin E$ **do**
  $p \leftarrow$ random number from $(0, 1)$
  if $p < \rho$ **then**
    $E \leftarrow E \cup \{\{v, w\}\}$
  end if
end for

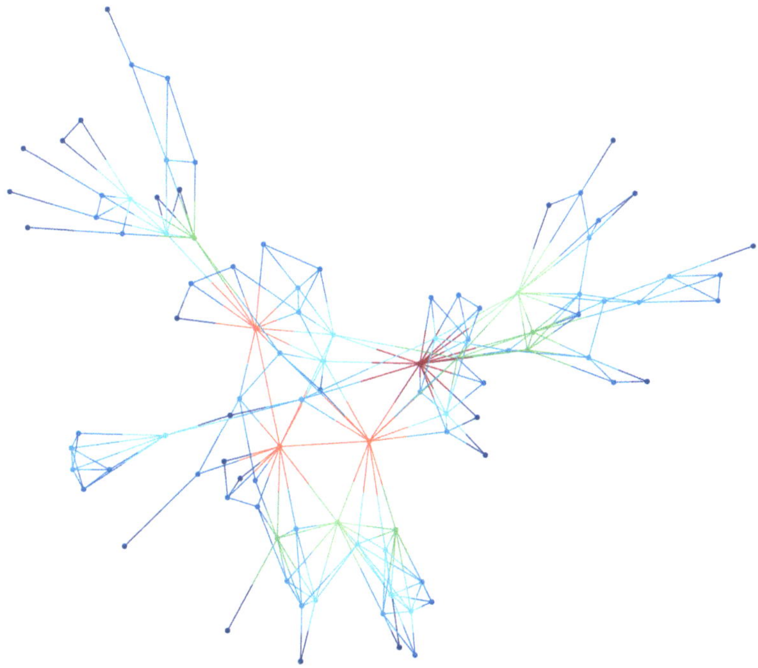

Fig. 3.6 A graph with 100 vertices generated by the duplication model from scratch. The model was run with parameter $\rho = 0.45$, which gives the exponent $\beta \approx 2.61$. The graph contains 246 edges, so that average degree is 4.92. If this graph represents a social network from 100 persons, each person has on average 5 friends

Used directly, the algorithm does allow to build a scale-free network, but we don't really want to use it this way. This is a typical *growth-only algorithm without rewiring*—vertices are added at every step (but not deleted) and the edges that were created previously are always kept. Let's add a few changes and see what happens to the network.

Firstly, we want to create several generations of agents. The inflow of new persons in LEMS happens in waves, and we want to take that into account in our model. We will need to slightly modify Algorithm 3.1 for that. In a simple modified version, there are k generations of size n added sequentially; the model is, thus, named $\mathcal{D}(\rho, k, n)$. Each vertex has two sub-networks: a coeval sub-network and a mature sub-network. The *coeval sub-network* is created together with the new generation of vertices by running $\mathcal{D}(\rho)$ n times. Then this new graph is added as a sub-graph to the overall graph and, for each vertex from the new generation, the vertex and edge duplication steps are performed. New vertices, however, are connected only to older generation vertices during this process—that's why this sub-network is called *mature sub-network*. Algorithm 3.2 formalises this process.

The graph that results from executing this algorithm with parameters $\mathcal{D}(0.45, 5, 20)$ is shown in Fig. 3.7. Compare it with Fig. 3.6. Both graphs have

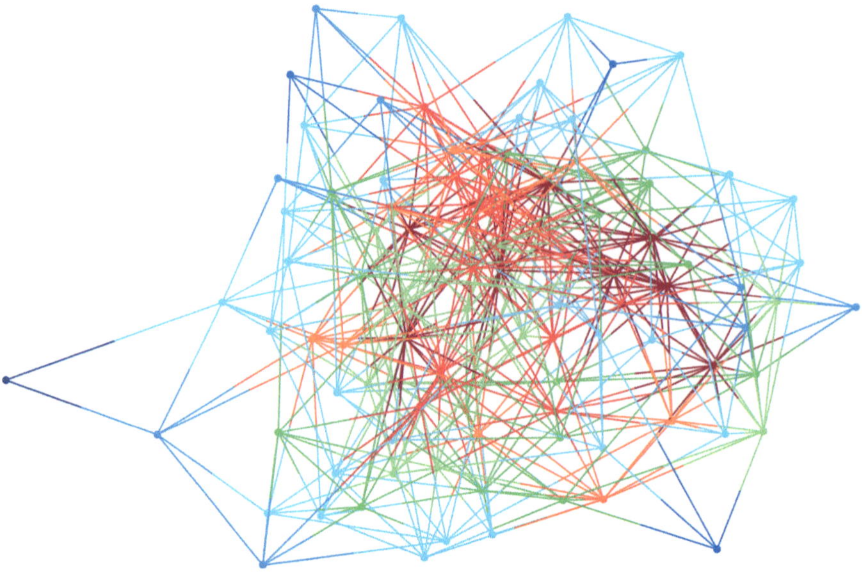

Fig. 3.7 A graph with 100 vertices generated by five generations forming coeval and mature networks. Created by Algorithm 3.2. Duplication model parameter was $\rho = 0.45$. Each generation of 20 vertices formed the coeval network (between themselves) using $\mathcal{D}(\rho)$. Then the generation formed the mature network with vertices from previous generations already present in the graph, also using $\mathcal{D}(\rho)$. The graph contains 525 edges, so that average degree is 10.5

Algorithm 3.2 Duplication model with k generations of size n creating coeval and mature networks $\mathcal{D}(\rho, k, n)$

$G \leftarrow$ empty graph
for $i = 1$ **to** k **do**
  $g \leftarrow$ empty graph
  {Creating coeval network:}
  for $j = 1$ **to** n **do**
    execute $\mathcal{D}(\rho)$ on g
  end for
  $G \leftarrow G \cup g$
  {Creating mature network for newly added vertices:}
  for all $v \in g.V$ **do**
    perform vertex and edge duplication, connecting with $u : u \in G.V$ **and** $u \notin g.V$
  end for
end for

100 vertices, the only difference is that the graph in Fig. 3.6 was created assuming one large generation, while the one in Fig. 3.7 was created assuming five generations of size 20 added sequentially and forming coeval and mature networks. The average degree of a vertex doubled as a result of this seemingly small change.

Secondly, if we are not interested in modelling very short-term events, we are typically interested in allowing agents to "die"—or exit the model. In the LEMS

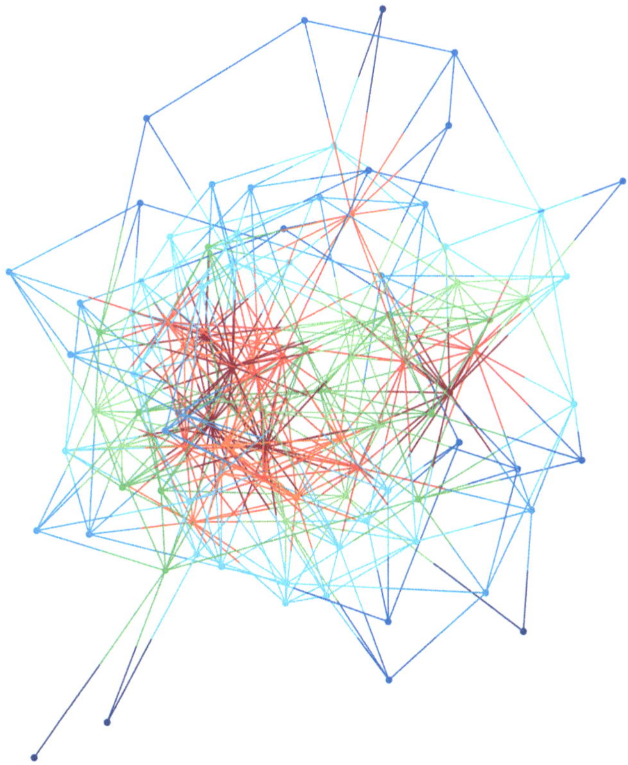

Fig. 3.8 A graph with 100 vertices generated by 100 generations forming coeval and mature networks with a lifespan of 5 periods. Created by Algorithm 3.2 modified to allow the deletion of "too old" vertex generations. Duplication model parameter was $\rho = 0.45$. Each generation of 20 vertices formed the coeval network (between themselves) using $\mathcal{D}(\rho)$. Then the generation formed the mature network with persons from previous generations already present in the graph, also using $\mathcal{D}(\rho)$. Generations are removed from the model five periods after being added. The graph contains 444 edges, so that average degree is 8.88

context, this can be equivalent to retirement. So far, the algorithms we used did not allow to delete vertices (and the incident edges) from the graph. To incorporate the ability to exit from the model, we have to add another modification: vertex generations are removed from the graph after "living" in the system for m generations (or periods). The model, thus, becomes $\mathcal{D}(\rho, k, n, m)$. The modification is straightforward, so it's not formalised in another algorithm. Figure 3.8 shows the graph generated with model $\mathcal{D}(0.45, 100, 20, 5)$; 100 generations of size 20 were generated, each remaining in the model for five periods. The parameters are set so that the resulting graph has 100 vertices, as the previous two graphs. The graph has a visually similar structure to Fig. 3.7, but average degree is in between those of graphs in Figs. 3.6 and 3.7.

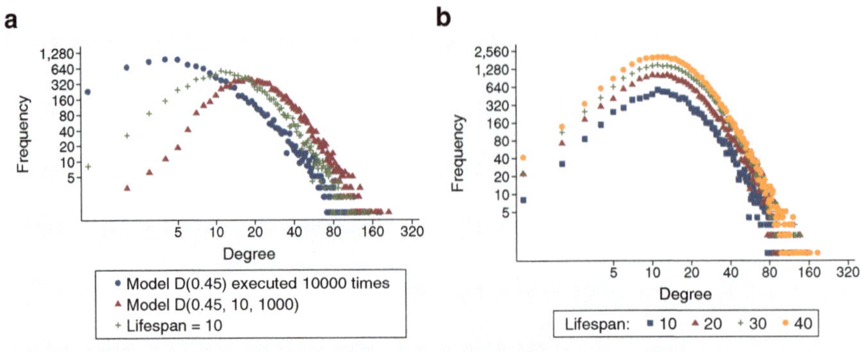

Fig. 3.9 (**a**) Degree distribution: comparison of Algorithms 3.1 and 3.2 to restricting lifespan to ten periods. Algorithm 3.1 was executed 10,000 times. Algorithm 3.2 created 10 generations sized 1000. For lifespan = 10, 100 generations were added. In all cases, $\rho = 0.45$. (**b**) Degree distribution in Algorithm 3.2 with limited lifespan: comparison of four lifespans. In all cases, $\rho = 0.45$ and $n = 1000$. As increasing lifespan increases graph size, 100 generations were added for lifespan = 10, 110 for lifespan = 20, 120 for lifespan = 30 and 130 for lifespan = 40

To sum up, after making two small modifications to the initial algorithm, we got a denser graph with an increase in average degree of around two times. A more interesting question would be, how these modifications affect the *distribution* of vertex degrees.

Figures 3.9a, b show a comparison of degree distributions. The size of graphs in these figures is at least 10,000, so that there are enough observations to see a smooth distribution. All graphs analysed in Fig. 3.9a have exactly 10,000 vertices. With increasing lifespan, however, the size of the graph increases together with lifespan: if every period, a generation of size n is added and exists in the model for m periods, the graph has nm vertices. Thus, graphs in Fig. 3.9b have different sizes (I consider keeping n constant across graphs more important than keeping nm constant). Moreover, m periods are needed for the graph to reach this size if it is built from scratch. For comparison, I allow all graphs to exist for the same number of periods after reaching the size nm—hence different values of parameter k, as noted under the figure.

As the figures show, degree distribution keeps its scale-free nature: in all graphs, it is linear (or close to linear) in log–log scale. However, the exponent β of the power law increases with more complex algorithms used. The slope of the linear part of the distribution is steeper in the graph generated by Algorithm 3.2 than in the graph generated by Algorithm 3.1 (see Fig. 3.9a). Moreover, the steepness (and, hence, the exponent β) increases with lifespan (see Fig. 3.9b).

Degree distribution is not the only characteristic of the generated network we are interested in. Figure 3.10 shows a comparison of average degree and clustering coefficient (or transitivity). The average degree is nearly three times higher in Algorithm 3.2 than in Algorithm 3.1 (with the parameters used), while

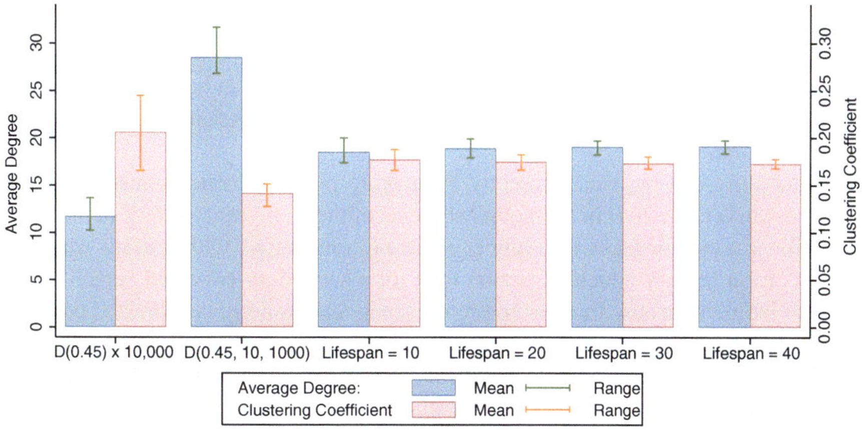

Fig. 3.10 Comparison of average degree and clustering coefficient of six graphs

the clustering coefficient is one-third lower. When the lifetime of vertices is limited, both characteristics are between the two extremes. Their values do not change substantially with lifespan, however.

3.3.2 *Defining Tie Creation and Destruction Principles*

We just saw two principles of tie creation and destruction that could be used in an agent-based model of LEMS: (1) when entering the model (e.g., after graduating secondary school), create a coeval and a mature sub-networks and (2) destroy ties when leaving the model (dying/retiring). These are basic principles and, clearly, they can be enriched. The enrichment strategy can be built around the mechanisms of homophily and economic usefulness.

Homophily can be used in a model directly and indirectly. A direct use is when there are several types of persons, and a person of type *A* creates connections either exclusively with type-*A* persons [257] or preferring them to other persons [1]. The set-up in [257], thus, leads to a network with two homogeneous connected components,[12] one for each type of persons, which may not necessarily be what you want in your model. The level of homophily-related probability of preference in [1] is different across agents and also changes over time, depending on social

[12]A *connected component* is a subgraph of an undirected graph where each pair of vertices is connected by a path and there are no other vertices in the graph that are connected to any vertex of the connected component but aren't in its vertex set. Thus, the connected components of a graph are isolated from each other—there is no edge between a vertex in one component and a vertex in any other component.

network, co-workers and average homophily in the society. If different job types (or professions) are modelled, persons could be constrained to create ties with the employees of their job type or their profession only [264] (or with a preference to creating ties with them). The direct use of homophily may also be more elaborate—e.g., social networks in [80] are generated to attain a pre-set level of *inbreeding homophily index.*[13]

Homophily can be used indirectly by making persons create connections with their co-workers [1, 264] or with other students of their university [262]. In the latter case, for instance, it leads to a higher share of connections with tertiary-educated persons for a tertiary-educated person than for a secondary-educated person. I call this use indirect because there is no preference as such to tertiary-educated persons, but the result is the same as if this preference was modelled directly. Formally, this use can be viewed as an implementation of the structural constraint mechanism defined in Chap. 1.

The economic usefulness mechanism sometimes appears to be motivated by homophily (e.g., in [117]), but I'd prefer to separate these two mechanisms. The difference is that homophily may be thought of as a realisation of individual's comfort in keeping ties with certain people ("I'm glad to be his friend"), while economic usefulness is a realisation of the economic utility of keeping these ties ("It pays to be his friend"). Typically, this mechanism is used in models that include referral hiring, and then creating ties with employed persons with a higher probability than with the unemployed [46, 117] reflects higher usefulness of employed neighbours in the social graph. That's because in case of unemployment or on-the-job search, chances to learn about a vacancy or receive an employment offer (depending on model specifics) are higher. For similar reasons, the economic usefulness mechanism is applied to destroying ties: connections with the unemployed have higher probability to be broken [46, 117]. Alternatively, the list of direct neighbours in the social network can be ordered descending by unemployment spell length (which is zero for the employed) and then the required number of ties are destroyed, starting with the beginning of this list (i.e., with those who have been unemployed for the longest time) [264].

Tie destruction is one of the ways how vertex degrees can be kept in a reasonable range. Of course, there are individuals with thousands of connections in today's online social networks, but in LEMS models, we typically want to prevent such large degrees. This can be done by imposing a ceiling on the number of connections—absolute number of connections or the number of new connections—or force the person to break a certain number of connections when new connections are made [264]. Again, the control over vertex degrees can be more elaborate. For instance, in

[13] *Inbreeding homophily of type i* is defined as $IH_i \equiv (H_i - w_i)/(1 - w_i)$, where H_i is the share of ties of type-i individuals with other type-i individuals and w_i is the share of type-i individuals in the population [77]. It measures the preference to creating ties with same-type individuals relative to the maximum bias, which is $(1 - w_i)$. The *inbreeding homophily index* is then $IH \equiv \sum_i w_i \times IH_i$ [80].

[117], every edge has a cost to its incident vertices and persons have to balance the total cost of their social ties with the likelihood of being referred when looking for a job (which increases with the number of ties). They do it by choosing a strategy from the set of strategies defining vertex degree using experience-weighted attraction algorithm [58, 59]. The use of this algorithm means that persons *learn* from their previous actions and, thus, adapt to the environment.[14]

3.3.3 Calibrating the Network Structure

After having decided to use social networks in the model and understanding how to generate networks with different structures, the next question is to decide which of these structures should be used in a particular model. There are two views of this issue: a "field" view and a "lab" view.

Proponents of the "lab" view hold the opinion that for several reasons (e.g., time or resource constraints or a dubious effectiveness of the "field" view), it is more efficient to experiment with different network structures until the model generates system behaviour that is in line with what is empirically observed. In other words, network structure in this case is an unknown parameter that has to be calibrated, hence its drawbacks. Besides increasing calibration parameter space, there may be problems if, e.g., the model has been created to simulate the effects of a certain policy and different network structures pass validation but have significantly different policy implications. Then policy-makers would ask a reasonable question: which of these networks corresponds to reality better?

Proponents of the "field" view would claim that the structure of the social network should be based on what is empirically observed when making the modelled decision. This sounds appealing, but also expensive and not necessarily useful. It may require a large-scale empirical study, which would take substantial time to execute and also require larger financial support than standard cross-sectional economic studies. Even if time and resources make such study possible, the methodology has to be carefully thought through, because in many contexts, during interviews or surveys people may not be telling about the true effects of social networks on their decisions. It may not necessarily be because they want to impress others by appearing to make decisions independently or rationally—it may easily be that they simply do not realise whether and how they used their social networks.

In this case, it's better to take a middle, "hybrid" view. Ideally, we want our agent-based model with social networks to be *cross-validated* in the sense of Moss–Edmonds [196]: agent micro-level behaviour should be validated *qualitatively* by

[14]Discussing learning and adaptation algorithms is beyond the scope of this book. You can find a short introduction to this field specifically for economic modelling in, e.g., [48]. For more details, see books on artificial intelligence.

domain experts, while system macro-level behaviour should be validated against available statistics.

Thus, ideally, the first step would be gathering information from available studies and from experts about how exactly social networks are formed in the precise context you're modelling. What's the average size of the network? Are there hierarchical layers depending on the frequency of contact? Which layers are likely to be more important in affecting individual decisions you're modelling? Are there actually several overlapping networks, each used for their own purposes (in other words, do individuals use different well-defined subsets of connections in different contexts)? How exactly do the homophily and economic usefulness mechanisms affect network formation? I wrote "ideally" before this list to indicate that it might be quite difficult to find concrete answers on many of these questions. In the worst case, you might have to get involved in an empirical study—but it will, most likely, not be as large-scale study as you would have to do had you not studied the published fieldwork results or contacted with field experts.

Then, after this qualitative information is incorporated in the model—more precisely, in the (evolutionary) algorithm that should generate the social network with the required characteristics (primarily, degree distribution), for instance, in the Duplication model or one of its variants mentioned above—there will be several parameters that cannot be set from qualitative data. For instance, it might be the maximum number of new co-workers with which the new employee creates ties. These will have to be calibrated to the values at which the model generates plausible macro-level characteristics of the network (e.g., average degree and clustering coefficient) and the system overall (e.g., education level attainment rates or the Beveridge curve).

3.4 To Embed or Not to Embed?

So you created your LEMS model featuring education market, labour market, persons whose decisions and actions are influenced by social networks and, perhaps, even firms. The model serves your purposes and generates you important results. Does your work end here? Not necessarily.

The model can be left as is, and there is nothing wrong with that. Any model, philosophically speaking, has a goal and a corresponding level of abstraction. It may certainly be the case that for a given goal, it is reasonable to assume that the larger context is exogenous (or fixed) and only the model itself should be studied. For instance, if the goal is to analyse how the process of choosing whether to enter higher education affects the overall rate of overeducation, then the level of abstraction might or might not require analysis at the level of fields of study. In any case, it does not perhaps make much sense to incorporate this model into a more general model of choosing education track starting from secondary school, unless inter-generational mobility of education is important for the goal of the model.

Nevertheless, some would argue that if you include firms in the model and set a probabilistic labour demand mechanism but don't include production and consumption sides, your system remains open. Your counter-arguments that adding non-primitive representation of production and consumption to the model will boost its complexity might not sound convincing to them. They would claim that because the system is not closed, the flow of finite resources is not controlled and, thus, the system may end up in a state in which it would hardly end up in reality.

Does it mean that you now have to think what your firms produce (and, of course, add firms first if you only have vacancies in your model), how they price it, who are consumers, what is their consumption behaviour? All that becomes even more interesting if you go beyond the assumption of producing a single homogeneous good. But then a firm may lack financing, so why not add a bank, together with competing banks and a central bank to control financial institutions? And then add a government that has to control everything and pay unemployment benefits and pensions? Sounds inspiring, right? If yes, then you can move on to embedding your model into a larger context typically relevant for macroeconomic studies.

Generally speaking, it makes sense to incorporate the given model into a larger context and study the resulting system in the following cases. Firstly, and primarily, it is when you have to evaluate important interrelationships between the given model and other components of a greater system. Typically, however, you do it after setting another goal of the model, which naturally requires the development of a new model, a sub-case of which is the re-use of an existing model to extend it or incorporate it into a larger model.

Secondly, it may be a successful smaller model that is ready to be incorporated as a sub-component into a larger, already developed, model. For instance, there might be a full macroeconomic model where the LEMS component lacked detail. After additional elaboration and analysis of its dynamics separately from other components, it may be included if it allows for greater precision in the results of the overall model.

Figure 3.11 shows how the resulting model may look, based on several existing macroeconomic agent-based models. Besides individuals, education institutions and firms, the model should have government (to collect taxes, pay unemployment benefits, employ persons in the public sector and issue public debt), the financial sector consisting of commercial banks (issuing credit to firms, providing deposits and other investment services to individuals and buying public debt) and the central bank (to regulate commercial banks, provide them credit and deposit services and also buy public debt if needed). Furthermore, firms may be divided into consumption good firms (actually producing the goods individuals would buy) and investment good firms (involved in research and development and operating in the B2B sector, selling the results of their research to consumption good firms as new capital investments).

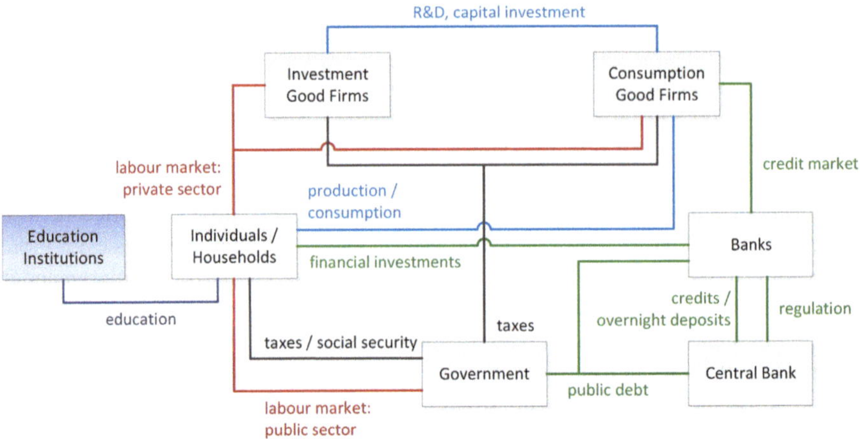

Fig. 3.11 Typical structure of a macroeconomic model with LEMS as a subsystem. *Source:* developed based on [83, 97, 219]. The Education Institutions segment is *highlighted*, as it is not included in the studies mentioned in the source statement of the figure, but it should be included as part of the LEMS subsystem

Concluding Remarks

If your first reaction after reading this book is that you want to know more about social networks in general and in the context of LEMS, in particular, then my goal has been reached. This is a fascinating topic and, while knowing some details about how networks work, there's still much space to dig into.

Many issues important in applying ABM to modelling LEMS were left out of scope or only dipped into, which should not be a surprise, given the size of this book. These include such topics as the construction of a vacancy; setting up the model and defining agent decisions related to overeducation; modelling the production decisions of firms; modelling the networks of firms; network statistics other than degree distribution, average degree and clustering coefficient; comparing the structure of two (an empirical and a generated) networks; modelling the job satisfaction of an agent; modelling learning/adaptation behaviours. The details behind most of these issues, however, can be found in the papers referenced in this book.

I expect further development of social network modelling (in the LEMS context) using ABM to go in the following non-exclusive directions:

- *Superimposition of several types of networks, where each sub-network plays a separate role.* We typically view an individual's network as homogeneous, in the sense that he always uses all his connections if necessary. The hierarchically layered structure of networks has been proved empirically, but are there sub-hierarchies or sub-networks? What are they, how are they created and how are they structured? For which (LEMS-related) issues are individuals using different sub-networks? Does accounting for the existence and use of sub-networks allow us to model macro-level dynamics (in LEMS) more accurately?
- *Plummeting importance of general social network generation algorithms.* These algorithms may be good for analytic results on social networks, but their details may not necessarily be valid in the contexts where they are applied. As our knowledge on the micro-dynamics in real social networks advances, agent-based models will increasingly incorporate the validated interaction mechanisms lead-

A. Tarvid, *Agent-Based Modelling of Social Networks in Labour–Education Market System*, SpringerBriefs in Complexity, DOI 10.1007/978-3-319-26539-1

ing to the creation of plausible networks, rather than general network generation algorithms. This tendency will be similar to economic modelling, where ABM allows to create models that are much closer to reality than analytically tractable mathematical models.

- *Increasing interest in applying ABM to practical issues related to LEMS policy.* In the LEMS context, mismatch between the employee and the job (like overeducation, for instance) is an important concern in empirical labour economics. ABM could provide substantial help in modelling the decision-making processes leading to mismatch and the policy context that can influence both the labour-market side and the education side of LEMS, helping to increase the share of matches in the labour market. There are, of course, many other issues of relevance to policy-makers in LEMS, such as increasing productivity and competitiveness or boosting the number of innovations and the return from them, and ABM is well applicable—and should be applied—there, too.
- *A shift of the focus to the entrepreneurial side of LEMS.* A lack of innovations, on the one hand, and unimpressive condition of the labour market, on the other, increase the interest in understanding the decisions to create firms. At the same time, this mechanism is insufficiently developed in agent-based models. This gap will certainly be closed in the future.

References

1. Abdou, M., Gilbert, N.: Modelling the emergence and dynamics of social and workplace segregation. Mind Soc. **8**(2), 173–191 (2009). doi:10.1007/s11299-009-0056-3
2. Adamic, L., Adar, E.: How to search a social network. Soc. Netw. **27**(3), 187–203 (2005). doi:10.1016/j.socnet.2005.01.007
3. Agut, S., Peiró, J.M., Grau, R.: The effect of overeducation on job content innovation and career-enhancing strategies among young Spanish employees. J. Career Dev. **36**(2), 159–182 (2009). doi:10.1177/0894845308327738
4. Ahituv, A., Lerman, R.I.: Job turnover, wage rates, and marital stability: how are they related? Rev. Econ. Househ. **9**(2), 221–249 (2011). doi:10.1007/s11150-010-9101-6
5. Ahn, Y.Y., Han, S., Kwak, H., Moon, S., Jeong, H.: Analysis of topological characteristics of huge online social networking services. In: Proceedings of the 16th International Conference on World Wide Web, WWW '07, pp. 835–844. ACM, New York (2007). doi:10.1145/1242572.1242685
6. Akgün, A.E., Keskin, H., Byrne, J.C., Ilhan, Z.: Complex adaptive system mechanisms, adaptive management practices, and firm product innovativeness. R&D Manag. **44**(1), 18–41 (2014). doi:10.1111/radm.12036
7. Alam, S., Ali Abbas, S., Edmonds, B.: Validating simulated networks: some lessons learned. In: Alam, S.J., Parunak, H.V.D. (eds.) Multi-Agent-Based Simulation XIV. Lecture Notes in Computer Science, vol. 8235, pp. 71–82. Springer, Berlin/Heidelberg (2014). doi:10.1007/978-3-642-54783-6_5
8. Albert, R., Barabási, A.L.: Statistical mechanics of complex networks. Rev. Mod. Phys. **74**, 47–97 (2002). doi:10.1103/RevModPhys.74.47
9. Aleksynska, M., Tritah, A.: Occupation–education mismatch of immigrant workers in Europe: context and policies. Econ. Educ. Rev. **36**, 229–244 (2013). doi:10.1016/j.econedurev.2013.06.001
10. Alexander, J.A., Lichtenstein, R., Oh, H.J., Ullman, E.: A causal model of voluntary turnover among nursing personnel in long-term psychiatric settings. Res. Nurs. Health **21**(5), 415–427 (1998). doi:10.1002/(SICI)1098-240X(199810)21:5<415::AID-NUR5>3.0.CO;2-Q
11. Allen, J., van der Velden, R.: Educational mismatches versus skill mismatches: effects on wages, job satisfaction, and on-the-job search. Oxf. Econ. Pap. **53**(3), 434–452 (2001). doi:10.1093/oep/53.3.434
12. Allen, J.A., Mueller, S.L.: The revolving door: a closer look at major factors in volunteers' intention to quit. J. Community Psychol. **41**(2), 139–155 (2013). doi:10.1002/jcop.21519

A. Tarvid, *Agent-Based Modelling of Social Networks in Labour–Education Market System*, SpringerBriefs in Complexity, DOI 10.1007/978-3-319-26539-1

13. Allisey, A.F., Noblet, A.J., Lamontagne, A.D., Houdmont, J.: Testing a model of officer intentions to quit: the mediating effects of job stress and job satisfaction. Crim. Just. Behav. **41**(6), 751–771 (2014). doi:10.1177/0093854813509987
14. Axtell, R.L.: Why agents? On the varied motivations for agent computing in the social sciences. Working paper 17, Center on Social and Economic Dynamics, The Brookings Institution (2000)
15. Axtell, R.: Effects of interaction topology and activation regime in several multi-agent systems. In: Moss, S., Davidsson, P. (eds.) Multi-Agent-Based Simulation. Lecture Notes in Computer Science, vol. 1979, pp. 33–48. Springer, Berlin/Heidelberg (2001). doi:10.1007/3-540-44561-7_3
16. Axtell, R.L.: What economic agents do: how cognition and interaction lead to emergence and complexity. Rev. Austrian Econ. **20**(2–3), 105–122 (2007). doi:10.1007/s11138-007-0021-5
17. Baert, S., Verhaest, D.: Unemployment or overeducation: which is a worse signal to employers? Discussion paper no. 8312, IZA (2014)
18. Baert, S., Cockx, B., Verhaest, D.: Overeducation at the start of the career: stepping stone or trap? Labour Econ. **25**, 123–140 (2013). doi:10.1016/j.labeco.2013.04.013
19. Bakker, A.B., Demerouti, E.: The job demands–resources model: state of the art. J. Manag. Psychol. **22**(3), 309–328 (2007). doi:10.1108/02683940710733115
20. Ballot, G.: Modeling the labor market as an evolving institution: model ARTEMIS. J. Econ. Behav. Organ. **49**(1), 51–77 (2002). doi:10.1016/S0167-2681(02)00058-6
21. Ballot, G., Kant, J.D., Goudet, O.: Modeling both sides of the French labor market with adaptive agents under bounded rationality. In: EAEPE Annual Conference, Paris (2013)
22. Bankes, S.C.: Agent-based modeling: a revolution? Proc. Natl. Acad. Sci. **99**(Suppl. 3), 7199–7200 (2002). doi:10.1073/pnas.072081299
23. Barabási, A.L.L., Bonabeau, E.: Scale-free networks. Sci. Am. **288**(5), 60–69 (2003)
24. Barone, C., Ortiz, L.: Overeducation among European university graduates: a comparative analysis of its incidence and the importance of higher education differentiation. High. Educ. **61**(3), 325–337 (2011). doi:10.1007/s10734-010-9380-0
25. Batty, M.: Cities and Complexity: Understanding Cities with Cellular Automata, Agent-Based Models, and Fractals. MIT Press, Cambridge (2005)
26. Bauer, T.K.: Educational mismatch and wages: a panel analysis. Econ. Educ. Rev. **21**(3), 221–229 (2002). doi:10.1016/S0272-7757(01)00004-8
27. Büchel, F.: Overqualification: reasons, measurement issues, and typological affinity to unemployment. In: Descy, P., Tessaring, M. (eds.) Training and Learning for Competence. Second Report on Vocational Training Research in Europe: Synthesis Report, vol. 2, pp. 453–560. Office for Official Publications of the European Communities, Luxembourg (2001)
28. Büchel, F.: The effects of overeducation on productivity in Germany – the firms' viewpoint. Econ. Educ. Rev. **21**(3), 263–275 (2002). doi:10.1016/S0272-7757(01)00020-6
29. Büchel, F., van Ham, M.: Overeducation, regional labor markets, and spatial flexibility. J. Urban Econ. **53**(3), 482–493 (2003). doi:10.1016/S0094-1190(03)00008-1
30. Becker, G.S.: Human Capital: A Theoretical and Empirical Analysis, with Special Reference to Education, 3rd edn. The University of Chicago Press, Chicago (1993)
31. Beckhusen, J., Florax, R.J., Poot, J., Waldorf, B.S.: Attracting global talent and then what? Overeducated immigrants in the United States. J. Reg. Sci. **53**(5), 834–854 (2013). doi:10.1111/jors.12030
32. Beinhocker, E.D.: The Origin of Wealth: Evolution, Complexity, and the Radical Remaking of Economics. Harvard Business School Press, Boston (2006)
33. Betti, G., D'Agostino, A., Neri, L.: Educational mismatch of graduates: a multidimensional and fuzzy indicator. Soc. Indic. Res. **103**(3), 465–480 (2011). doi:10.1007/s11205-010-9712-6
34. Bewley, T.F.: Why Wages Don't Fall During a Recession. Harvard University Press, Cambridge (1999)
35. Bianchi, C., Cirillo, P., Gallegati, M., Vagliasindi, P.A.: Validating and calibrating agent-based models: a case study. Comput. Econ. **30**(3), 245–264 (2007). doi:10.1007/s10614-007-9097-z

36. Birkin, M., Wu, B.: A review of microsimulation and hybrid agent-based approaches. In: Heppenstall, A.J., Crooks, A.T., See, L.M., Batty, M. (eds.) Agent-Based Models of Geographical Systems, pp. 51–68. Springer, Dordrecht (2012). doi:10.1007/978-90-481-8927-4_3
37. Blázquez, M., Budría, S.: Overeducation dynamics and personality. Educ. Econ. **20**(3), 260–283 (2012). doi:10.1080/09645292.2012.679338
38. Blenkinsopp, J., Scurry, T.: "Hey GRINGO!": the HR challenge of graduates in non-graduate occupations. Pers. Rev. **36**(4), 623–637 (2007). doi:10.1108/00483480710752830
39. Boccaletti, S., Latora, V., Moreno, Y., Chavez, M., Hwang, D.U.: Complex networks: structure and dynamics. Phys. Rep. **424**(4–5), 175–308 (2006). doi:10.1016/j.physrep.2005.10.009
40. Boudarbat, B.: Field of study choice by community college students in Canada. Econ. Educ. Rev. **27**(1), 79–93 (2008). doi:10.1016/j.econedurev.2006.06.001
41. Boudarbat, B., Montmarquette, C.: Choice of fields of study of university Canadian graduates: the role of gender and their parents' education. Educ. Econ. **17**(2), 185–213 (2009). doi:10.1080/09645290802133032
42. Boudon, R.: Education, Opportunity, and Social Inequality: Changing Prospects in Western Society. Wiley, New York (1974)
43. Bourdieu, P.: Distinction: A Social Critique of the Judgement of Taste. Harvard University Press, Boston (1984)
44. Bourdieu, P., Passeron, J.C.: Reproduction in Education, Society and Culture. Sage, London (1990)
45. Bracke, P., Pattyn, E., von dem Knesebeck, O.: Overeducation and depressive symptoms: diminishing mental health returns to education. Sociol. Health Illn. **35**(8), 1242–1259 (2013). doi:10.1111/1467-9566.12039
46. Bramoullé, Y., Saint-Paul, G.: Social networks and labor market transitions. Labour Econ. **17**(1), 188–195 (2010). doi:10.1016/j.labeco.2009.09.011
47. Breen, R., Goldthorpe, J.H.: Explaining educational differentials: towards a formal rational action theory. Ration. Soc. **9**(3), 275–305 (1997). doi:10.1177/104346397009003002
48. Brenner, T.: Agent learning representation: advice on modelling economic learning. In: Tesfatsion, L., Judd, K.L. (eds.) Handbook of Computational Economics, Chap. 18, vol. 2, pp. 895–947. Elsevier, Amsterdam (2006). doi:10.1016/S1574-0021(05)02018-6
49. Brenner, T., Werker, C.: A taxonomy of inference in simulation models. Comput. Econ. **30**(3), 227–244 (2007). doi:10.1007/s10614-007-9102-6
50. Brooks, B., Hogan, B., Ellison, N., Lampe, C., Vitak, J.: Assessing structural correlates to social capital in Facebook ego networks. Soc. Netw. **38**, 1–15 (2014). doi:10.1016/j.socnet.2014.01.002
51. Brough, P., Frame, R.: Predicting police job satisfaction and turnover intentions: the role of social support and police organisational variables. N. Z. J. Psychol. **33**(1), 8–16 (2004)
52. Brown, M., Setren, E., Topa, G.: Do informal referrals lead to better matches? Evidence from a firm's employee referral system. Discussion paper no. 8175, IZA (2014)
53. Budría, S.: Are educational mismatches responsible for the 'inequality increasing effect' of education? Soc. Indic. Res. **102**(3), 409–437 (2011). doi:10.1007/s11205-010-9675-7
54. Byrne, M., Flood, B., Hassall, T., Joyce, J., Montaño, J.L.A., González, J.M.G., Tourna-Germanou, E.: Motivations, expectations and preparedness for higher education: a study of accounting students in Ireland, the UK, Spain and Greece. Account. Forum **36**(2), 134–144 (2012). doi:10.1016/j.accfor.2011.12.001
55. Cai, Y.: Choosing fields of graduate study among Chinese and Taiwanese students. Soc. Sci. J. **40**(3), 495–501 (2003). doi:10.1016/S0362-3319(03)00047-8
56. Cainarca, G.C., Sgobbi, F.: The return to education and skills in Italy. Int. J. Manpow. **33**(2), 187–205 (2012). doi:10.1108/01437721211225444
57. Caliendo, M., Schmidl, R., Uhlendorff, A.: Social networks, job search methods and reservation wages: evidence for Germany. Int. J. Manpow. **32**(7), 796–824 (2011). doi:10.1108/01437721111174767

58. Camerer, C., Ho, T.H.: Experience-weighted attraction learning in coordination games: probability rules, heterogeneity, and time-variation. J. Math. Psychol. **42**(2–3), 305–326 (1998). doi:10.1006/jmps.1998.1217
59. Camerer, C., Ho, T.H.: Experienced-weighted attraction learning in normal form games. Econometrica **67**(4), 827–874 (1999)
60. Campbell, D.P., Borgen, F.H., Eastes, S.H., Johansson, C.B., Peterson, R.A.: A set of basic interest scales for the strong vocational interest blank for men. J. Appl. Psychol. **52**(6), 1–54 (1968). doi:10.1037/h0026495
61. Cappellari, L., Tatsiramos, K.: With a little help from my friends? Quality of social networks, job finding and job match quality. Eur. Econ. Rev. **78**, 55–75 (2015). doi:10.1016/j.euroecorev.2015.04.002
62. Carroll, D., Tani, M.: Over-education of recent higher education graduates: new Australian panel evidence. Econ. Educ. Rev. **32**, 207–218 (2013). doi:10.1016/j.econedurev.2012.10.002
63. Chevalier, A.: Measuring over-education. Economica **70**(279), 509–531 (2003). doi:10.1111/1468-0335.t01-1-00296
64. Chevalier, A., Lindley, J.: Overeducation and the skills of UK graduates. J. R. Stat. Soc. Ser. A Stat. Soc. **172**(2), 307–337 (2009). doi:10.1111/j.1467-985X.2008.00578.x
65. Chiu, T.H.Y., Chen, C.C., Joung, Y.J., Chen, S.: A study of blog networks to determine online social network properties from the tie strength perspective. Online Inf. Rev. **38**(3), 381–398 (2014). doi:10.1108/OIR-01-2013-0022
66. Chow, I.H.S., Ng, I., Gong, Y.Y.: Risk-taking and relational perspective on turnover intentions. Int. J. Hum. Resour. Manag. **23**(4), 779–792 (2012). doi:10.1080/09585192.2011.561249
67. Chung, F., Lu, L.: Complex Graphs and Networks. Regional Conference Series in Mathematics, vol. 107. American Mathematical Society, Providence (2006)
68. Colman, A.M.: A Dictionary of Psychology, 4th edn. Oxford University Press, Oxford (2015)
69. Comer, K.W.: Who goes first? An examination of the impact of activation on outcome behaviour in agent-based models. Ph.D. thesis, George Mason University (2014)
70. Comer, K.W., Loerch, A.G.: The impact of agent activation on population behavior in an agent-based model of civil revolt. Procedia Comput. Sci. **20**, 183–188 (2013). doi:10.1016/j.procs.2013.09.258
71. Cortese, C.G., Colombo, L., Ghislieri, C.: Determinants of nurses' job satisfaction: the role of work–family conflict, job demand, emotional charge and social support. J. Nurs. Manag. **18**(1), 35–43 (2010). doi:10.1111/j.1365-2834.2009.01064.x
72. Costa, P.T., McCrae, R.R.: Revised NEO Personality Inventory (NEO-PI-R) and NEO Five-Factor Inventory (NEO-FFI) Manual. Psychological Assessment Resources, Odessa (1992)
73. Cottini, E., Kato, T., Westergaard-Nielsen, N.: Adverse workplace conditions, high-involvement work practices and labor turnover: evidence from Danish linked employer–employee data. Labour Econ. **18**(6), 872–880 (2011). doi:10.1016/j.labeco.2011.07.003
74. Cranny, C.J., Smith, P.C., Stone, E.F.: Job Satisfaction: How People Feel About Their Jobs and How It Affects Their Performance. Lexington Press, New York (1992)
75. Croce, G., Ghignoni, E.: Demand and supply of skilled labour and overeducation in Europe: a country-level analysis. Comp. Econ. Stud. **54**(2), 413–439 (2012). doi:10.1057/ces.2012.12
76. Crossley, C.D., Bennett, R.J., Jex, S.M., Burnfield, J.L.: Development of a global measure of job embeddedness and integration into a traditional model of voluntary turnover. J. Appl. Psychol. **92**(4), 1031–1042 (2007). doi:10.1037/0021-9010.92.4.1031
77. Currarini, S., Jackson, M.O., Pin, P.: An economic model of friendship: homophily, minorities, and segregation. Econometrica **77**(4), 1003–1045 (2009). doi:10.3982/ECTA7528
78. Cutillo, A., di Pietro, G.: The effects of overeducation on wages in Italy: a bivariate selectivity approach. Int. J. Manpow. **27**(2), 143–168 (2006). doi:10.1108/01437720610666182
79. Dar, Y., Getz, S.: Learning ability, socioeconomic status, and student placement for undergraduate studies in Israel. High. Educ. **54**(1), 41–60 (2007). doi:10.1007/s10734-006-9045-1
80. Dawid, H., Gemkow, S.: How do social networks contribute to wage inequality? Insights from an agent-based analysis. Ind. Corp. Change **23**(5), 1171–1200 (2014). doi:10.1093/icc/dtt049

81. Dawid, H., Gemkow, S., Harting, P., Kabus, K., Wersching, K., Neugart, M.: Skills, innovation, and growth: an agent-based policy analysis. J. Econ. Stat. **228**(2–3), 251–275 (2008)
82. Dawid, H., Gemkow, S., Harting, P., Neugart, M.: On the effects of skill upgrading in the presence of spatial labor market frictions: an agent-based analysis of spatial policy design. J. Artif. Soc. Soc. Simul. **12**(4), 5 (2009)
83. Dawid, H., Gemkow, S., Harting, P., van der Hoog, S., Neugart, M.: The Eurace@ Unibi model: an agent-based macroeconomic model for economic policy analysis. Working papers in Economics and Management 05-2012, Bielefeld University (2012)
84. Dawid, H., Gemkow, S., Harting, P., Neugart, M.: Labor market integration policies and the convergence of regions: the role of skills and technology diffusion. J. Evol. Econ. **22**(3), 543–562 (2012). doi:10.1007/s00191-011-0245-1
85. Dawid, H., Harting, P., Neugart, M.: Spatial labor market frictions and economic convergence: policy implications from a heterogeneous agent model. Working papers in Economics and Management 09-2013, Bielefeld University (2013)
86. Deckert, A., Klein, R.: Agentenbasierte Simulation zur Analyse und Lösung betriebswirtschaftlicher Entscheidungsprobleme. J. Betriebswirt. **60**(2), 89–125 (2010). doi:10.1007/s11301-010-0058-6
87. Deissenberg, C., van der Hoog, S., Dawid, H.: EURACE: a massively parallel agent-based model of the European economy. Appl. Math. Comput. **204**(2), 541–552 (2008). doi:10.1016/j.amc.2008.05.116
88. Dekker, R., de Grip, A., Heijke, H.: The effects of training and overeducation on career mobility in a segmented labour market. Int. J. Manpow. **23**(2), 106–125 (2002). doi:10.1108/01437720210428379
89. Demerouti, E., Bakker, A.B., Nachreiner, F., Shaufeli, W.B.: The job demands–resources model of burnout. J. Appl. Psychol. **86**(3), 499–512 (2001). doi:10.1037/0021-9010.86.3.499
90. de Oliveira, M.M., Santos, M., Kiker, B.: The role of human capital and technological change in overeducation. Econ. Educ. Rev. **19**(2), 199–206 (2000). doi:10.1016/S0272-7757(99)00020-5
91. Derycke, H., Vlerick, P., Burnay, N., Decleire, C., D'Hoore, W., Hasselhorn, H.M., Braeckman, L.: Impact of the effort–reward imbalance model on intent to leave among Belgian health care workers: a prospective study. J. Occup. Organ. Psychol. **83**(4), 879–893 (2010). doi:10.1348/096317909X477594
92. DG Employment, Social Affairs and Inclusion: employment and social developments in Europe 2012. Tech. rep., European Commission (2012)
93. Di Pietro, G.: Technological change, labor markets, and 'low-skill, low-technology traps'. Technol. Forecast. Soc. Chang. **69**(9), 885–895 (2002). doi:10.1016/S0040-1625(01)00182-2
94. Dickey, H., Watson, V., Zangelidis, A.: Job satisfaction and quit intentions of offshore workers in the UK North Sea oil and gas industry. Scott. J. Polit. Econ. **58**(5), 607–633 (2011). doi:10.1111/j.1467-9485.2011.00561.x
95. Diem, A., Wolter, S.C.: Overeducation among Swiss university graduates: determinants and consequences. J. Labour Mark. Res. **47**(4), 313–328 (2014). doi:10.1007/s12651-014-0164-3
96. Dorogovtsev, S.N., Mendes, J.F.F.: Evolution of networks. Adv. Phys. **51**(4), 1079–1187 (2002). doi:10.1080/00018730110112519
97. Dosi, G., Fagiolo, G., Napoletano, M., Roventini, A.: Income distribution, credit and fiscal policies in an agent-based Keynesian model. J. Econ. Dyn. Control **37**(8), 1598–1625 (2013). doi:10.1016/j.jedc.2012.11.008
98. Ducharme, L.J., Martin, J.K.: Unrewarding work, coworker support, and job satisfaction: a test of the buffering hypothesis. Work Occup. **27**(2), 223–243 (2000). doi:10.1177/0730888400027002005
99. Dunbar, R.I.: Coevolution of neocortical size, group size and language in humans. Behav. Brain Sci. **16**(4), 681–694 (1993). doi:10.1017/S0140525X00032325
100. Dunbar, R.I., Arnaboldi, V., Conti, M., Passarella, A.: The structure of online social networks mirrors those in the offline world. Soc. Netw. **43**, 39–47 (2015). doi:10.1016/j.socnet.2015.04.005

101. Ellis, B.: An overview of complexity theory: understanding primary care as a complex adaptive system. In: Sturmberg, J.P., Martin, C.M. (eds.) Handbook of Systems and Complexity in Health, pp. 485–494. Springer, New York (2013). doi:10.1007/978-1-4614-4998-0_29
102. Ellis, B., Herbert, S.I.: Complex adaptive systems (cas): an overview of key elements, characteristics and application to management theory. Inform. Prim. Care **19**(1), 33–37 (2011)
103. Epstein, J.M., Axtell, R.: Growing Artificial Societies: Social Science from the Bottom Up. The Brookings Institution, Washington (1996)
104. Ertas, N.: Turnover intentions and work motivations of millennial employees in federal service. Public Personnel Manage. (2015). doi:10.1177/0091026015588193
105. Fagiolo, G., Dosi, G., Gabriele, R.: Matching, bargaining, and wage setting in an evolutionary model of labor market and output dynamics. Adv. Complex Syst. **7**(2), 157–186 (2004). doi:10.1142/S0219525904000135
106. Fagiolo, G., Moneta, A., Windrum, P.: A critical guide to empirical validation of agent-based models in economics: methodologies, procedures, and open problems. Comput. Econ. **30**(3), 195–226 (2007). doi:10.1007/s10614-007-9104-4
107. Felps, W., Mitchell, T.R., Hekman, D.R., Lee, T.W., Holtom, B.C., Harman, W.S.: Turnover contagion: how coworkers' job embeddedness and job search behaviors influence quitting. Acad. Manag. J. **52**(3), 545–561 (2009). doi:10.5465/AMJ.2009.41331075
108. Fernández, C., Ortega, C.: Labor market assimilation of immigrants in Spain: employment at the expense of bad job-matches? Span. Econ. Rev. **10**(2), 83–107 (2008). doi:10.1007/s10108-007-9032-4
109. Figari, F., Paulus, A., Sutherland, H.: Microsimulation and policy analysis. In: Atkinson, A.B., Bourguignon, F. (eds.) Handbook of Income Distribution. Handbook of Income Distribution, Chap. 24, vol. 2, pp. 2141–2221. Elsevier, Amsterdam (2015). doi:10.1016/B978-0-444-59429-7.00025-X
110. Firth, L., Mellor, D.J., Moore, K.A., Loquet, C.: How can managers reduce employee intention to quit? J. Manag. Psychol. **19**(2), 170–187 (2004). doi:10.1108/02683940410526127
111. Fischer, C.S.: To Dwell Among Friends: Personal Networks in Town and City. The University of Chicago Press, Chicago (1982)
112. Frank, R.H.: Why women earn less: the theory and estimation of differential overqualification. Am. Econ. Rev. **68**(3), 360–373 (1978)
113. Franzen, A., Hangartner, D.: Social networks and labour market outcomes: the non-monetary benefits of social capital. Eur. Sociol. Rev. **22**(4), 353–368 (2006). doi:10.1093/esr/jcl001
114. Freeman, J.A., Hirsch, B.T.: College majors and the knowledge content of jobs. Econ. Educ. Rev. **27**(5), 517–535 (2008). doi:10.1016/j.econedurev.2007.07.004
115. Frei, C., Sousa-Poza, A.: Overqualification: permanent or transitory? Appl. Econ. **44**(14), 1837–1847 (2012). doi:10.1080/00036846.2011.554380
116. Frenette, M.: The overqualified Canadian graduate: the role of the academic program in the incidence, persistence, and economic returns to overqualification. Econ. Educ. Rev. **23**(1), 29–45 (2004). doi:10.1016/S0272-7757(03)00043-8
117. Gemkow, S., Neugart, M.: Referral hiring, endogenous social networks, and inequality: an agent-based analysis. J. Evol. Econ. **21**(4), 703–719 (2011). doi:10.1007/s00191-011-0219-3
118. Ghignoni, E., Verashchagina, A.: Educational qualifications mismatch in Europe. Is it demand or supply driven? J. Comp. Econ. **42**(3), 670–692 (2014). doi:10.1016/j.jce.2013.06.006
119. Ghosh, D., Gurunathan, L.: Linking perceived corporate social responsibility and intention to quit: the mediating role of job embeddedness. Vis. J. Bus. Perspect. **18**(3), 175–183 (2014). doi:10.1177/0972262914539219
120. Gonçalves, B., Perra, N., Vespignani, A.: Modeling users' activity on Twitter networks: validation of Dunbar's number. PLoS One **6**(8), e22656 (2011). doi:10.1371/journal.pone.0022656
121. Goyette, K.A., Mullen, A.L.: Who studies the arts and sciences? Social background and the choice and consequences of undergraduate field of study. J. High. Educ. **77**(3), 497–538 (2006)

122. Granovetter, M.: The impact of social structure on economic outcomes. J. Econ. Perspect. **19**(1), 33–50 (2005). doi:10.1257/0895330053147958
123. Griffeth, R.W., Hom, P.W., Gaertner, S.: A meta-analysis of antecedents and correlates of employee turnover: update, moderator tests, and research implications for the next millennium. J. Manag. **26**(3), 463–488 (2000). doi:10.1177/014920630002600305
124. Groeneveld, S., Hartog, J.: Overeducation, wages and promotions within the firm. Labour Econ. **11**(6), 701–714 (2004). doi:10.1016/j.labeco.2003.11.005
125. Grossetti, M.: Where do social relations come from?: a study of personal networks in the Toulouse area of France. Soc. Netw. **27**(4), 289–300 (2005). doi:10.1016/j.socnet.2004.11.004
126. Gu, X., Blackmore, K.L.: A systematic review of agent-based modelling and simulation applications in the higher education domain. High. Educ. Res. Dev. 1–16 (2015). doi:10.1080/07294360.2015.1011088
127. Guimerà, R., Danon, L., Díaz-Guilera, A., Giralt, F., Arenas, A.: Self-similar community structure in a network of human interactions. Phys. Rev. E **68**, 065103 (2003). doi:10.1103/PhysRevE.68.065103
128. Guironnet, J.P., Peypoch, N.: Human capital allocation and overeducation: a measure of French productivity (1987, 1999). Econ. Model. **24**(3), 398–410 (2007). doi:10.1016/j.econmod.2006.09.003
129. Harland, K., Heppenstall, A.J.: Using agent-based models for education planning: is the UK education system agent based? In: Heppenstall, A.J., Crooks, A.T., See, L.M., Batty, M. (eds.) Agent-Based Models of Geographical Systems, pp. 481–497. Springer, Dordrecht (2012). doi:10.1007/978-90-481-8927-4_23
130. Hartog, J.: Over-education and earnings: where are we, where should we go? Econ. Educ. Rev. **19**(2), 131–147 (2000). doi:10.1016/S0272-7757(99)00050-3
131. Hausknecht, J.P., Trevor, C.O.: Collective turnover at the group, unit, and organizational levels: evidence, issues, and implications. J. Manag. **37**(1), 352–388 (2011). doi:10.1177/0149206310383910
132. Hayes, L.J., O'Brien-Pallas, L., Duffield, C., Shamian, J., Buchan, J., Hughes, F., Laschinger, H.K.S., North, N.: Nurse turnover: a literature review—an update. Int. J. Nurs. Stud. **49**(7), 887–905 (2012). doi:10.1016/j.ijnurstu.2011.10.001
133. Hazans, M.: Labor market integration of ethnic minorities in Latvia. In: Kahanec, M., Zimmerman, K.F. (eds.) Ethnic Diversity in European Labor Markets: Challenges and Solutions, pp. 163–197. Edward Elgar, Cheltenham (2011)
134. Hegselmann, R., Flache, A.: Understanding complex social dynamics: a plea for cellular automata based modelling. J. Artif. Soc. Soc. Simul. **1**(3), 1 (1998)
135. Henrickson, L.: Old wine in a new wineskin: college choice, college access using agent-based modeling. Soc. Sci. Comput. Rev. **20**(4), 400–419 (2002). doi:10.1177/089443902237319
136. Hill, R.A., Dunbar, R.I.: Social network size in humans. Hum. Nat. **14**(1), 53–72 (2003). doi:10.1007/s12110-003-1016-y
137. Hobfoll, S.E.: Conservation of resources: a new attempt at conceptualizing stress. Am. Psychol. **44**(3), 513–524 (1989). doi:10.1037/0003-066X.44.3.513
138. Hoff, K., Stiglitz, J.E.: Modern economic theory and development. In: Meier, G.M., Stiglitz, J.E. (eds.) Frontiers of Development Economics: The Future in Perspective, pp. 389–459. Oxford University Press, New York (2001)
139. Holland, J.H.: Studying complex adaptive systems. J. Syst. Sci. Complex. **19**(1), 1–8 (2006). doi:10.1007/s11424-006-0001-z
140. Holland, J.H., Miller, J.H.: Artificial adaptive agents in economic theory. Am. Econ. Rev. **81**(2), 365–370 (1991)
141. Holland, J.L.: A theory of vocational choice. J. Couns. Psychol. **6**(1), 35–45 (1959). doi:10.1037/h0040767
142. Holtom, B.C., Mitchell, T.R., Lee, T.W., Eberly, M.B.: Turnover and retention research: a glance at the past, a closer review of the present, and a venture into the future. Acad. Manag. Ann. **2**(1), 231–274 (2008). doi:10.1080/19416520802211552

143. Hu, N.B.: Effects of college students' perceptions of labor market variables and conditions on their choice of academic majors. In: The 36th Annual Conference of The Association for Institutional Research (1996)
144. ILO: Global employment trends for youth 2013: a generation at risk. Report, International Labour Office, Geneva (2013)
145. Jauhiainen, S.: Overeducation in the Finnish regional labour markets. Pap. Reg. Sci. **90**(3), 573–588 (2011). doi:10.1111/j.1435-5957.2010.00334.x
146. Jensen, U., Gartner, H., Rässler, S.: Estimating German overqualification with stochastic earnings frontiers. Adv. Stat. Anal. **94**(1), 33–51 (2010). doi:10.1007/s10182-009-0121-6
147. Jiang, K., Liu, D., McKay, P.F., Lee, T.W., Mitchell, T.R.: When and how is job embeddedness predictive of turnover? A meta-analytic investigation. J. Appl. Psychol. **97**(5), 1077–1096 (2012). doi:10.1037/a0028610
148. Johnson, J.V., Hall, E.M.: Job strain, work place social support, and cardiovascular disease: a cross-sectional study of a random sample of the Swedish working population. Am. J. Public Health **78**(10), 1336–1342 (1988). doi:10.2105/AJPH.78.10.1336
149. Jones, M.K., Jones, R.J., Latreille, P.L., Sloane, P.J.: Training, job satisfaction, and workplace performance in Britain: evidence from WERS 2004. Labour **23**, 139–175 (2009). doi:10.1111/j.1467-9914.2008.00434.x
150. Judge, T.A., Thoresen, C.J., Bono, J.E., Patton, G.K.: The job satisfaction–job performance relationship: a qualitative and quantitative review. Psychol. Bull. **127**(3), 376–407 (2001). doi:10.1037/0033-2909.127.3.376
151. Kalleberg, A.L.: Work values and job rewards: a theory of job satisfaction. Am. Sociol. Rev. **42**(1), 124–143 (1977)
152. Karakaya, G., Plasman, R., Rycx, F.: Overeducation on the Belgian labour market: evaluation and analysis of the explanatory factors through two types of approaches. Comp. J. Comp. Int. Educ. **37**(4), 513–532 (2007). doi:10.1080/03057920701366317
153. Karasek, R., Theorell, T.: Healthy Work: Stress, Productivity, and the Reconstruction of Working Life. Basic Books, New York (1990)
154. Kember, D., Hong, C., Ho, A., Ho, A.: More can mean less motivation: applying a motivational orientation framework to the expanded entry into higher education in Hong Kong. Stud. High. Educ. **36**(2), 209–225 (2011). doi:10.1080/03075070903514062
155. Kennedy, M.: A review of system dynamics models of educational policy issues. In: 27th International Conference of the System Dynamics Society 2009, vol. 3, pp. 1661–1683 (2009)
156. Kiersztyn, A.: Stuck in a mismatch? The persistence of overeducation during twenty years of the post-communist transition in Poland. Econ. Educ. Rev. **32**, 78–91 (2013). doi:10.1016/j.econedurev.2012.09.009
157. Kirman, A.P.: The intrinsic limits of modern economic theory: the emperor has no clothes. Econ. J. **99**(395), 126–139 (1989)
158. Kirman, A.P.: Whom or what does the representative individual represent? J. Econ. Perspect. **6**(2), 117–136 (1992)
159. Korpi, T., Tåhlin, M.: Educational mismatch, wages, and wage growth: overeducation in Sweden, 1974–2000. Labour Econ. **16**(2), 183–193 (2009). doi:10.1016/j.labeco.2008.08.004
160. Koster, F., de Grip, A., Fouarge, D.: Does perceived support in employee development affect personnel turnover? Int. J. Hum. Resour. Manag. **22**(11), 2403–2418 (2011). doi:10.1080/09585192.2011.584404
161. Kramarz, F., Skans, O.N.: When strong ties are strong: networks and youth labour market entry. Rev. Econ. Stud. **81**(3), 1164–1200 (2014). doi:10.1093/restud/rdt049
162. Kuddo, A.: Employment services and active labor market programs in Eastern European and Central Asian countries. SP Discussion paper 0918, The World Bank (2009)
163. Kumar, R., Novak, J., Tomkins, A.: Structure and evolution of online social networks. In: Yu, P.S., Han, J., Faloutsos, C. (eds.) Link Mining: Models, Algorithms, and Applications, pp. 337–357. Springer, New York (2010). doi:10.1007/978-1-4419-6515-8_13
164. Kwak, H., Lee, C., Park, H., Moon, S.: What is Twitter, a social network or a news media? In: Proceedings of the 19th International Conference on World Wide Web, WWW '10, pp. 591–600. ACM, New York (2010). doi:10.1145/1772690.1772751

165. Lee, G., Jimenez, B.S.: Does performance management affect job turnover intention in the Federal government? Am. Rev. Public Adm. **41**(2), 168–184 (2011). doi:10.1177/0275074010368991
166. Leijonhufvud, A.: Limits to the equilibrating capabilities of market systems. J. Econ. Interac. Coord. **4**(2), 173–182 (2009). doi:10.1007/s11403-009-0052-z
167. Lemieux, T.: Occupations, fields of study and returns to education. Can. J. Econ. Rev. Can. Econ. **47**(4), 1047–1077 (2014). doi:10.1111/caje.12116
168. Leombruni, R., Richiardi, M.: LABORsim: an agent-based microsimulation of labour supply – an application to Italy. Comput. Econ. **27**(1), 63–88 (2006). doi:10.1007/s10614-005-9016-0
169. Leontaridi, R.M., Ward, M.E.: Work-related stress, quitting intentions and absenteeism. Discussion paper no. 493, IZA (2002)
170. Leskovec, J., Horvitz, E.: Planetary-scale views on a large instant-messaging network. In: Proceedings of the 17th International Conference on World Wide Web, WWW '08, pp. 915–924. ACM, New York (2008). doi:10.1145/1367497.1367620
171. Levin, S.A.: Ecosystems and the biosphere as complex adaptive systems. Ecosystems **1**(5), 431–436 (1998). doi:10.1007/s100219900037
172. Levin, S., Xepapadeas, T., Crépin, A.S., Norberg, J., de Zeeuw, A., Folke, C., Hughes, T., Arrow, K., Barrett, S., Daily, G., Ehrlich, P., Kautsky, N., Mäler, K.G., Polasky, S., Troell, M., Vincent, J.R., Walker, B.: Social-ecological systems as complex adaptive systems: modeling and policy implications. Environ. Dev. Econ. **18**, 111–132 (2013). doi:10.1017/S1355770X12000460
173. Lewkovicz, Z., Thiriot, S., Caillou, P.: How detailed should social networks be for labor market's models? In: Social Networks and Multiagent Systems Symposium, York (2011)
174. Li, J., O'Donoghue, C.: A survey of dynamic microsimulation models: uses, model structure and methodology. Int. J. Microsimul. **6**(2), 3–55 (2013)
175. Lianos, T.P., Asteriou, D., Agiomirgianakis, G.: Foreign university graduates in the Greek labour market: employment, salaries and overeducation. Int. J. Finance Econ. **9**(2), 151–164 (2004). doi:10.1002/ijfe.238
176. Lis, G.A., Hanson, P., Burgermeister, D., Banfield, B.: Transforming graduate nursing education in the context of complex adaptive systems: implications for master's and {DNP} curricula. J. Prof. Nurs. **30**(6), 456–462 (2014). doi:10.1016/j.profnurs.2014.05.003
177. Locke, E.A.: What is job satisfaction? Organ. Behav. Hum. Perform. **4**(4), 309–336 (1969). doi:10.1016/0030-5073(69)90013-0
178. Lorscheid, I., Meyer, M., Hocke, S.: Simulation model and data analysis: where are we and where should we go? In: ESSA 2013: 9th Conference of the European Social Simulation Association (2013)
179. Luuger, G.F.: Artificial Intelligence: Structures and Strategies for Complex Problem Solving. Pearson Education, Boston (2009)
180. Maroulis, S., Bakshy, E., Gomez, L., Wilensky, U.: Modeling the transition to public school choice. J. Artif. Soc. Soc. Simul. **17**(2), 3 (2014)
181. Martin, C.W., Neugart, M.: Shocks and endogenous institutions: an agent-based model of labor market performance in turbulent times. Comput. Econ. **33**(1), 31–46 (2009). doi:10.1007/s10614-008-9149-z
182. Mason, M. (ed.): Complexity Theory and the Philosophy of Education. Wiley, Chichester (2008)
183. Mavromaras, K., McGuinness, S., O'Leary, N., Sloane, P., Fok, Y.K.: The problem of overskilling in Australia and Britain. Manch. Sch. **78**(3), 219–241 (2010). doi:10.1111/j.1467-9957.2009.02136.x
184. McCrae, R.R., Costa, P.T.: Introduction to the empirical and theoretical status of the five-factor model of personality traits. In: Widiger, T.A., Costa, P.T. (eds.) Personality Disorders and the Five-Factor Model of Personality, 3rd edn., pp. 15–27. American Psychological Association, Washington (2013)

185. McGee, S., Edson, R.: Challenges of governance in complex adaptive systems: a case study of U.S. public education. Procedia Comput. Sci. **36**, 131–139 (2014). doi:10.1016/j.procs.2014.09.049
186. McGuinness, S., Bennett, J.: Overeducation in the graduate labour market: a quantile regression approach. Econ. Educ. Rev. **26**(5), 521–531 (2007). doi:10.1016/j.econedurev.2005.12.003
187. McGuinness, S., Wooden, M.: Overskilling, job insecurity, and career mobility. Ind. Relat. J. Econ. Soc. **48**(2), 265–286 (2009). doi:10.1111/j.1468-232X.2009.00557.x
188. McPherson, M., Smith-Lovin, L., Cook, J.M.: Birds of a feather: homophily in social networks. Annu. Rev. Sociol. **27**, 415–444 (2001). doi:10.1146/annurev.soc.27.1.415
189. Mitchell, T.R., Holtom, B.C., Lee, T.W., Sablynski, C.J., Erez, M.: Why people stay: using job embeddedness to predict voluntary turnover. Acad. Manag. J. **44**(6), 1102–1121 (2001)
190. Mitzenmacher, M.: A brief history of generative models for power law and lognormal distributions. Internet Math. **1**(2), 226–251 (2004). doi:10.1080/15427951.2004.10129088
191. Mitzlaff, F., Atzmueller, M., Hotho, A., Stumme, G.: The social distributional hypothesis: a pragmatic proxy for homophily in online social networks. Soc. Netw. Anal. Min. **4**(1), 216 (2014). doi:10.1007/s13278-014-0216-2
192. Moniarou-Papaconstantinou, V., Tsatsaroni, A., Katsis Athanassios nad Koulaidis, V.: LIS as a field of study: socio-cultural influences on students' decision-making. ASLIB Proc. **62**(3), 321–344 (2010). doi:10.1108/00012531011046934
193. Montgomery, J.D.: Social networks and labor-market outcomes: toward an economic analysis. Am. Econ. Rev. **81**(5), 1408–1418 (1991)
194. Montmarquette, C., Cannings, K., Mahseredjian, S.: How do young people choose college majors? Econ. Educ. Rev. **21**(6), 543–556 (2002). doi:10.1016/S0272-7757(01)00054-1
195. Mora, J.G., García-Aracil, A., Vila, L.: Job satisfaction among young European higher education graduates. High. Educ. **53**(1), 29–59 (2007). doi:10.1007/s10734-005-2377-4
196. Moss, S., Edmonds, B.: Sociology and simulation: statistical and qualitative cross-validation. Am. J. Sociol. **110**(4), 1095–1131 (2005)
197. Moynihan, D.P., Pandey, S.K.: The ties that bind: social networks, person-organization value fit, and turnover intention. J. Public Adm. Res. Theory **18**(2), 205–227 (2008). doi:10.1093/jopart/mum013
198. Müller, W.: Education and youth integration into European labour markets. Int. J. Comp. Sociol. **46**(5–6), 461–485 (2005). doi:10.1177/0020715205060048
199. Murillo, I.P., Rahona-López, M., del Mar Salinas-Jiménez, M.: Effects of educational mismatch on private returns to education: an analysis of the Spanish case (1995–2006). J. Policy Model. **34**(5), 646–659 (2012). doi:10.1016/j.jpolmod.2011.07.012
200. Neugart, M.: Labor market policy evaluation with ACE. J. Econ. Behav. Organ. **67**(2), 418–430 (2008). doi:10.1016/j.jebo.2006.12.006
201. Newman, M.E.J.: Networks: An Introduction. Oxford University Press, New York (2010)
202. Ng, T.S., Kang, Y.C.: The inquiry of labor market dynamics in knowledge economies: an agent-based approach. IEEE Trans. Syst. Man Cybern. Syst. **43**(2), 416–427 (2013). doi:10.1109/TSMCA.2012.2195167
203. Nikolai, C., Madey, G.: Tools of the trade: a survey of various agent based modeling platforms. J. Artif. Soc. Soc. Simul. **12**(2), 2 (2009)
204. North, M.J., Macal, C.M.: Managing Business Complexity: Discovering Strategic Solutions with Agent-Based Modeling and Simulation. Oxford University Press, Oxford (2007)
205. Nyberg, A.J., Ployhart, R.E.: Context-emergent turnover (cet) theory: a theory of collective turnover. Acad. Manag. Rev. **38**(1), 109–131 (2013). doi:10.5465/amr.2011.0201
206. OECD: Education at a glance 2013: OECD indicators. Tech. rep., OECD Publishing (2013). doi:10.1787/eag-2013-en
207. Ohana, M., Meyer, M.: Should I stay or should I go now? Investigating the intention to quit of the permanent staff in social enterprises. Eur. Manag. J. **28**(6), 441–454 (2010). doi:10.1016/j.emj.2010.06.007

208. Ortiz, L., Kucel, A.: Do fields of study matter for over-education? The cases of Spain and Germany. Int. J. Comp. Sociol. **49**(4–5), 305–327 (2008). doi:10.1177/0020715208093079
209. Peiró, J.M., Agut, S., Grau, R.: The relationship between overeducation and job satisfaction among young Spanish workers: the role of salary, contract of employment, and work experience. J. Appl. Soc. Psychol. **40**(3), 666–689 (2010). doi:10.1111/j.1559-1816.2010.00592.x
210. Pike, G.R.: Students' personality types, intended majors, and college expectations: further evidence concerning psychological and sociological interpretations of Holland's theory. Res. High. Educ. **47**(7), 801–822 (2006). doi:10.1007/s11162-006-9016-5
211. Piracha, M., Tani, M., Vadean, F.: Immigrant over- and under-education: the role of home country labour market experience. IZA J. Migr. **1**(3) (2012). doi:10.1186/2193-9039-1-3
212. Pomazan, V., Mihalaşcu, D., Petcu, L.C., Gîrtu, M.A.: Enrolling in science and engineering academic programs—motivating and deterring factors. AIP Conf. Proc. **1203**(1), 1372–1377 (2010). doi:10.1063/1.3322374
213. Quintini, G.: Over-qualified or under-skilled: a review of existing literature. OECD Employment and Migration Working paper No. 121, OECD Publishing (2011). doi:10.1787/5kg58j9d7b6d-en
214. Railsback, S.F., Lytinen, S.L., Jackson, S.K.: Agent-based simulation platforms: review and development recommendations. Simulation **82**(9), 609–623 (2006). doi:10.1177/0037549706073695
215. Ralston, C.A., Borgen, F.H., Rottinghaus, P.J., Donnay, D.A.: Specificity in interest measurement: basic interest scales and major field of study. J. Vocat. Behav. **65**(2), 203–216 (2004). doi:10.1016/S0001-8791(03)00096-4
216. Ramos, R., Sanromá, E.: Overeducation and local labour markets in Spain. Tijdschrift voor economische en sociale geografie **104**(3), 278–291 (2013). doi:10.1111/j.1467-9663.2012.00752.x
217. Reardon, S.F., Kasman, M., Klasik, D., Baker, R.: Agent-based simulation models of the college sorting process. Working paper, Center for Education Policy Analysis, Stanford University (2014)
218. Reimer, D., Pollak, R.: Educational expansion and its consequences for vertical and horizontal inequalities in access to higher education in West Germany. Eur. Sociol. Rev. **26**(4), 415–430 (2010). doi:10.1093/esr/jcp029
219. Riccetti, L., Russo, A., Gallegati, M.: An agent based decentralized matching macroeconomic model. J. Econ. Interac. Coord. 1–28 (2014). doi:10.1007/s11403-014-0130-8
220. Richiardi, M.: Toward a non-equilibrium unemployment theory. Comput. Econ. **27**(1), 135–160 (2006). doi:10.1007/s10614-005-9019-x
221. Richiardi, M., Leombruni, R., Saam, N., Sonnessa, M.: A common protocol for agent-based social simulation. J. Artif. Soc. Soc. Simul. **9**(1), 15 (2006)
222. Robbins, S.P., Judge, T.A., Campbell, T.: Organizational Behaviour. Pearson Education Ltd., Harlow (2010)
223. Robinson, R.N., Kralj, A., Solnet, D.J., Goh, E., Callan, V.: Thinking job embeddedness not turnover: towards a better understanding of frontline hotel worker retention. Int. J. Hosp. Manag. **36**, 101–109 (2014). doi:10.1016/j.ijhm.2013.08.008
224. Robst, J.: Career mobility, job match, and overeducation. East. Econ. J. **21**(4), 539–550 (1995)
225. Robst, J.: Overeducation and college major: expanding the definition of mismatch between schooling and jobs. Manch. Sch. **76**(4), 349–368 (2008). doi:10.1111/j.1467-9957.2008.01064.x
226. Roxburgh, S.: Exploring the work and family relationship: gender differences in the influence of parenthood and social support on job satisfaction. J. Fam. Issues **20**(6), 771–788 (1999). doi:10.1177/019251399020006003
227. Rubb, S.: Overeducation: a short or long run phenomenon for individuals? Econ. Educ. Rev. **22**(4), 389–394 (2003). doi:10.1016/S0272-7757(02)00052-3
228. Rubb, S.: Overeducation in the labor market: a comment and re-analysis of a meta-analysis. Econ. Educ. Rev. **22**(6), 621–629 (2003). doi:10.1016/S0272-7757(02)00077-8

229. Rubinstein, G.: The Big Five among male and female students of different faculties. Personal. Individ. Differ. **38**(7), 1495–1503 (2005). doi:10.1016/j.paid.2004.09.012
230. Sakoda, J.M.: The checkerboard model of social interaction. J. Math. Sociol. **1**(1), 119–132 (1971). doi:10.1080/0022250X.1971.9989791
231. Santé, I., García, A.M., Miranda, D., Crecente, R.: Cellular automata models for the simulation of real-world urban processes: a review and analysis. Landsc. Urban Plan. **96**(2), 108–122 (2010). doi:10.1016/j.landurbplan.2010.03.001
232. Sattinger, M.: Assignment models of the distribution of earnings. J. Econ. Lit. **31**(2), 831–880 (1993)
233. Schelling, T.C.: Dynamic models of segregation. J. Math. Sociol. **1**(2), 143–186 (1971). doi:10.1080/0022250X.1971.9989794
234. Schleicher, D.J., Watt, J.D., Greguras, G.J.: Reexamining the job satisfaction-performance relationship: the complexity of attitudes. J. Appl. Psychol. **89**(1), 165–177 (2004)
235. Scott, A., Gravelle, H., Simoens, S., Bojke, C., Sibbald, B.: Job satisfaction and quitting intentions: a structural model of British general practitioners. Br. J. Ind. Relat. **44**(3), 519–540 (2006). doi:10.1111/j.1467-8543.2006.00511.x
236. Semmer, N.K., Elfering, A., Baillod, J., Berset, M., Beehr, T.A.: Push and pull motivations for quitting. Zeitschrift für Arbeits- und Organisationspsychologie A&O **58**(4), 173–185 (2014). doi:10.1026/0932-4089/a000167
237. Shields, M.A., Price, S.W.: Racial harassment, job satisfaction and intentions to quit: evidence from the British nursing profession. Economica **69**(274), 295–326 (2002). doi:10.1111/1468-0335.00284
238. Sicherman, N.: "Overeducation" in the labor market. J. Labor Econ. **9**(2), 101–122 (1991)
239. Siegrist, J.: Adverse health effects of high-effort/low-reward conditions. J. Occup. Health Psychol. **1**(1), 27–41 (1996). doi:10.1037/1076-8998.1.1.27
240. Silva, S.T., Valente, J.M., Teixeira, A.A.: An evolutionary model of industry dynamics and firms' institutional behavior with job search, bargaining and matching. J. Econ. Interac. Coord. **7**(1), 23–61 (2012). doi:10.1007/s11403-011-0085-y
241. Simões, C., Soares, A.M.: Applying to higher education: information sources and choice factors. Stud. High. Educ. **35**(4), 371–389 (2010). doi:10.1080/03075070903096490
242. Simon, H.A.: Models of Man, Social and Rational: Mathematical Essays on Rational Human Behavior in a Social Setting. Wiley, New York (1957)
243. Simon, H.A.: The Sciences of the Artificial, 3rd edn. The MIT Press, Cambridge (1996)
244. Skalli, A., Theodossiou, I., Vasileiou, E.: Jobs as Lancaster goods: facets of job satisfaction and overall job satisfaction. J. Socio Econ. **37**(5), 1906–1920 (2008). doi:10.1016/j.socec.2008.04.003
245. Smart, J.C., Feldman, K.A., Ethington, C.A.: Academic Disciplines: Holland's Theory and the Study of College Students and Faculty. Vanderbilt University Press, Nashville (2000)
246. Sousa-Poza, A., Henneberger, F.: Analyzing job mobility with job turnover intentions: an international comparative study. J. Econ. Issues **38**(1), 113–137 (2004)
247. Sparreboom, T., Tarvid, A.: Job polarization and skills mismatch in Europe (2016, forthcoming)
248. Spence, M.: Job market signaling. Q. J. Econ. **87**(3), 355–374 (1973). doi:10.2307/1882010
249. Steel, R.P., Lounsbury, J.W.: Turnover process models: review and synthesis of a conceptual literature. Hum. Resour. Manag. Rev. **19**(4), 271–282 (2009). doi:10.1016/j.hrmr.2009.04.002
250. Steinmetz, S., De Vries, D.H., Tijdens, K.G.: Should I stay or should I go? The impact of working time and wages on retention in the health workforce. Hum. Resour. Health **12**(4), 23 (2014). doi:10.1186/1478-4491-12-23
251. Stigler, G.J.: The economics of information. J. Polit. Econ. **69**(3), 213–225 (1961)
252. Stigler, G.J.: Information in the labor market. J. Polit. Econ. **70**(5), 94–105 (1962)
253. Støren, L.A., Wiers-Jenssen, J.: Foreign diploma versus immigrant background: determinants of labour market success or failure? J. Stud. Int. Educ. **14**(1), 29–49 (2010). doi:10.1177/1028315308327951

254. Strauss, L.M., Borenstein, D.: A system dynamics model for long-term planning of the undergraduate education in Brazil. High. Educ. **69**(3), 375–397 (2015). doi:10.1007/s10734-014-9781-6
255. Sutherland, J.: Qualifications mismatch and skills mismatch. Educ. Train. **54**(7), 619–632 (2012). doi:10.1108/00400911211265666
256. Takase, M.: A concept analysis of turnover intention: implications for nursing management. Collegian **17**(1), 3–12 (2010). doi:10.1016/j.colegn.2009.05.001
257. Takács, K., Squazzoni, F., Bravo, G.: The network antidote: an agent-based model of discrimination in labor markets. In: MKE 2012 Conference, Budapest (2012)
258. Tani, M.: Does immigration policy affect the education-occupation mismatch? Evidence from Australia. Aust. Bull. Labour **38**(2), 111–140 (2012)
259. Tarvid, A.: Effects of the field of higher education on the prevalence of over-education in European countries. In: Kesner, J. (ed.) Education: Evaluation, Reform and Policy, pp. 167–184. Athens Institute for Education and Research, Athens (2012)
260. Tarvid, A.: Field of study choice and migration: effects of social networks. In: Koch, A., Mandl, P. (eds.) Modeling Social Phenomena in Spatial Context. Geosimulation Series, vol. 2, pp. 125–138. LIT Publisher, Münster (2013)
261. Tarvid, A.: Unobserved heterogeneity in overeducation models: is personality more important than ability? Procedia Econ. Finance **5**, 722–731 (2013). doi:10.1016/S2212-5671(13)00084-1
262. Tarvid, A.: The effectiveness of access restriction to higher education in decreasing overeducation. Econ. Anal. Policy **45**, 11–26 (2015). doi:10.1016/j.eap.2014.12.003
263. Tarvid, A.: Job satisfaction determinants of tertiary graduates in Europe. Procedia Econ. Finance **24**, 682–690 (2015). doi:10.1016/S2212-5671(15)00674-7
264. Tarvid, A.: Job satisfaction as a unified mechanism for agent behaviour in a labour market with referral hiring. Int. J. Comput. Econ. Econ. (2016, forthcoming)
265. Tassier, T., Menczer, F.: Emerging small-world referral networks in evolutionary labor markets. IEEE Trans. Evol. Comput. **5**(5), 482–492 (2001). doi:10.1109/4235.956712
266. Tassier, T., Menczer, F.: Social network structure, segregation, and equality in a labor market with referral hiring. J. Econ. Behav. Organ. **66**(3–4), 514–528 (2008). doi:10.1016/j.jebo.2006.07.003
267. Tavares, O., Cardoso, S.: Enrolment choices in Portuguese higher education: do students behave as rational consumers? High. Educ. **66**(3), 297–309 (2013). doi:10.1007/s10734-012-9605-5
268. Tesfatsion, L.: Structure, behavior, and market power in an evolutionary labor market with adaptive search. J. Econ. Dyn. Control **25**(3–4), 419–457 (2001). doi:10.1016/S0165-1889(00)00032-4
269. The Economist: Agents of change. The Economist. http://www.economist.com/node/16636121 (July 2010)
270. Thiriot, S., Lewkovicz, Z., Caillou, P., Kant, J.D.: Referral hiring and labor markets: a computational study. In: Osinga, S., Hofstede, G.J., Verwaart, T. (eds.) Emergent Results of Artificial Economics. Lecture Notes in Economics and Mathematical Systems, vol. 652, pp. 15–25. Springer, Berlin/Heidelberg (2011). doi:10.1007/978-3-642-21108-9_2
271. Thurow, L.C.: Generating Inequality. Basic Books, New York (1975)
272. Tinbergen, J.: On the theory of income distribution. Weltwirtschaftliches Arch. **77**, 155–175 (1956)
273. Toivonen, R., Kovanen, L., Kivelä, M., Onnela, J.P., Saramäki, J., Kaski, K.: A comparative study of social network models: network evolution models and nodal attribute models. Soc. Netw. **31**(4), 240–254 (2009). doi:10.1016/j.socnet.2009.06.004
274. Tolsma, J., Need, A., de Jong, U.: Explaining participation differentials in Dutch higher education: the impact of subjective success probabilities on level choice and field choice. Eur. Sociol. Rev. **26**(2), 235–252 (2010). doi:10.1093/esr/jcp061
275. Torrens, P.M., O'Sullivan, D.: Cellular automata and urban simulation: where do we fo from here? Environ. Plan. B **28**(2), 163–168 (2001). doi:10.1068/b2802ed

276. Trevor, C.O., Nyberg, A.J.: Keeping your headcount when all about you are losing theirs: downsizing, voluntary turnover rates, and the moderating role of HR practices. Acad. Manag. J. **51**(2), 259–276 (2008). doi:10.5465/AMJ.2008.31767250
277. Tsai, Y.: Returns to overeducation: a longitudinal analysis of the U.S. labor market. Econ. Educ. Rev. **29**(4), 606–617 (2010). doi:10.1016/j.econedurev.2010.01.001
278. Uluhan, E., Aydin, M.N.: Complex adaptive systems theory in the context of business process management. In: Zehbold, C. (ed.) S-BPM ONE - Application Studies and Work in Progress. Communications in Computer and Information Science, vol. 422, pp. 147–156. Springer, Cham (2014). doi:10.1007/978-3-319-06191-7_10
279. Van Deuren, R., Santema, S.: How to choose your minor? Decision making variables used in the selection of a minor by undergraduate students from a Dutch university of applied sciences. Working papers 2012/06, Maastricht School of Management (2012)
280. Van der Hoog, S., Dawid, H.: Bubbles, crashes and the financial cycle: insights from a stock-flow consistent agent-based macroeconomic model. Working papers in Economics and Management 01-2015, Bielefeld University (2015)
281. Van de Werfhorst, H.G., de Graaf, N.D., Kraaykamp, G.: Intergenerational resemblance in field of study in the Netherlands. Eur. Sociol. Rev. **17**(3), 275–293 (2001). doi:10.1093/esr/17.3.275
282. Van de Werfhorst, H.G., Sullivan, A., Cheung, S.Y.: Social class, ability and choice of subject in secondary and tertiary education in Britain. Br. Educ. Res. J. **29**(1), 41–62 (2003). doi:10.1080/0141192032000057366
283. Verhaest, D., Omey, E.: The determinants of overeducation: different measures, different outcomes? Int. J. Manpow. **31**(6), 608–625 (2010). doi:10.1108/01437721011073337
284. Verhaest, D., Omey, E.: Overeducation, undereducation and earnings: further evidence on the importance of ability and measurement error bias. J. Labor Res. **33**(1), 76–90 (2012). doi:10.1007/s12122-011-9125-6
285. Verhaest, D., Van der Velden, R.: Cross-country differences in graduate overeducation. Eur. Sociol. Rev. **29**(3), 642–653 (2013). doi:10.1093/esr/jcs044
286. Vila, L.E., García-Aracil, A., Mora, J.G.: The distribution of job satisfaction among young European graduates: does the choice of study field matter? J. High. Educ. **78**(1), 97–118 (2007)
287. Visse, D., Coetzee, S.: Affective-cognitive consistency of attitude as a moderator of the job satisfaction-performance relationship. SA J. Ind. Psychol. **31**(3), 62–69 (2005)
288. Vossensteyn, J.J.H.: Perceptions of student price-responsiveness: a behavioural economics exploration of the relationships between socio-economic status, perceptions of financial incentives and student choice. Ph.D. thesis, University of Twente (2005)
289. Wald, S.: The impact of overqualification on job search. Int. J. Manpow. **26**(2), 140–156 (2005). doi:10.1108/01437720510597649
290. Walker, I., Zhu, Y.: Differences by degree: evidence of the net financial rates of return to undergraduate study for England and Wales. Econ. Educ. Rev. **30**(6), 1177–1186 (2011). doi:10.1016/j.econedurev.2011.01.002
291. Wang, G., Wang, B., Wang, T., Nika, A., Zheng, H., Zhao, B.Y.: Whispers in the dark: analysis of an anonymous social network. In: Proceedings of the 2014 Conference on Internet Measurement Conference, IMC '14, pp. 137–150. ACM, New York (2014). doi:10.1145/2663716.2663728
292. Weiss, H.M.: Deconstructing job satisfaction: separating evaluations, beliefs and affective experiences. Hum. Resour. Manag. Rev. **12**(2), 173–194 (2002). doi:10.1016/S1053-4822(02)00045-1
293. William Lee, T., Burch, T.C., Mitchell, T.R.: The story of why we stay: a review of job embeddedness. Annu. Rev. Organ. Psychol. Organ. Behav. **1**(1), 199–216 (2014). doi:10.1146/annurev-orgpsych-031413-091244
294. Wilson, C., Boe, B., Sala, A., Puttaswamy, K.P., Zhao, B.Y.: User interactions in social networks and their implications. In: Proceedings of the 4th ACM European conference on Computer systems, EuroSys '09, pp. 205–218. ACM, New York, NY (2009). doi:10.1145/1519065.1519089

295. Wirz, A., Atukeren, E.: An analysis of perceived overqualification in the Swiss labor market. Econ. Bull. **9**(2), 1–10 (2005)
296. World Economic Forum: Global Risks 2014, 9th edn. Insight Report, Geneva (2014)
297. Zafar, B.: Double majors: one for me, one for the parents? Econ. Inq. **50**(2), 287–308 (2012). doi:10.1111/j.1465-7295.2011.00403.x
298. Zheng, H., Son, Y.J., Chiu, Y.C., Head, L., Feng, Y., Xi, H., Kim, S., Hickman, M.: A Primer for Agent-Based Simulation and Modeling in Transportation Applications, Chap. 3. U.S. Department of Transportation (2013)
299. Zhou, W.X., Sornette, D., Hill, R.A., Dunbar, R.I.: Discrete hierarchical organization of social group sizes. Proc. R. Soc. Lond. B Biol. Sci. **272**(1561), 439–444 (2005). doi:10.1098/rspb.2004.2970
300. Zimmerman, R.D.: Understanding the impact of personality traits on individuals' turnover decisions: a meta-analytic path model. Pers. Psychol. **61**(2), 309–348 (2008). doi:10.1111/j.1744-6570.2008.00115.x

Index

A. Tarvid, *Agent-Based Modelling of Social Networks in Labour–Education Market System*, SpringerBriefs in Complexity, DOI 10.1007/978-3-319-26539-1

MIX
Papier aus verantwortungsvollen Quellen
Paper from responsible sources
FSC® C105338

If you have any concerns about our products,
you can contact us on
ProductSafety@springernature.com

In case Publisher is established outside the EU,
the EU authorized representative is:
Springer Nature Customer Service Center GmbH
Europaplatz 3, 69115 Heidelberg, Germany

Printed by Libri Plureos GmbH
in Hamburg, Germany